A Step-by-Step Approach to
ELEMENTARY
ORGANIC SYNTHESIS

M. Paul Servé

Associate Professor of Chemistry
Wright State University
Dayton, Ohio

ANN ARBOR SCIENCE
PUBLISHERS INC
P.O. BOX 1425 • ANN ARBOR, MICH. 48106

During the past two decades the amount of information gathered in the area of organic chemistry has increased tremendously. To see how this vast amount of knowledge has been reflected in elementary organic chemistry courses, one need only look at the organic chemistry texts which have appeared on the market in the last five years. Most of the new texts emphasize a core of reactions, mechanisms and structure determinations via spectral properties. These subjects are then supplemented with special topics such as biochemistry, polymer chemistry and natural products.

The subject of multi-step organic synthesis, however, is normally just touched in most organic texts. Nevertheless, problems regarding a knowledge of multi-step synthesis are frequently asked at the end of most chapters in the average organic book. It appears that the presentation of this subject is left entirely to the instructor to be covered either in the formal lecture portion of the course or during informal review sessions.

When I first began teaching organic chemistry, I found that while conducting review sessions a frequently asked question was "How do I do this synthesis?". In most cases the student did not even know where to begin. If the problem required the synthesis of a molecule from a specified starting material, the student would normally begin the synthesis with the starting material and try to reach the desired product. In most cases he would fail and become discouraged and demoralized.

About five years ago, I decided to make a package of synthesis problems which would illustrate some of the most useful synthetic procedures in organic chemistry. The first year proved to be of limited success with regard to facilitating the students' under-standing of organic chemistry. It was not until later, when I divided the synthesis into functional group chapters and wrote out in detail the worked syntheses of several compounds in each chapter, that I noticed the students comprehending much better the material being presented in the lecture. Many students began questioning the mechanistic pathways and stereochemical conse-quences of reactions which were involved in the synthesis prob-lems. Word spread and soon students from other organic classes came to my office and asked for any spare copies of my synthesis handouts. Several members of the faculty convinced me that I ought to combine the various organic synthesis handouts into a book which could be used as a supplement to an organic chem-istry text. This is that book.

The book is divided into chapters based on the preparation of individual functional groups. Each chapter contains a few reac-tions used to prepare a particular functional group, a worked syntheses and an unworked syntheses problem section. The worked syntheses section is set up in a question-answer type format. The syntheses of a molecule is worked out by taking the desired product and asking the question "How can this mole-cule be prepared in one step from another molecule A?". If A is not a permitted starting material, the question is then asked "How can A be prepared in one step from another molecule B?". This format is continued until a permitted starting material is arrived at.

This question-answer format serves several useful functions. It shows the student where he should begin each synthetic problem, namely with the desired product. It allows the student to always arrive at the correct product, since he is beginning the synthetic sequence with it. A complicated synthetic scheme containing several steps is broken down into a series of simple one-step reac-tions. The format compels the student to fully understand a reaction since he must know how to use it in a forward and backward manner. Success is attained in working problems and this in turn leads to encouragement and over-all success in learning organic chemistry.

It should be noted that the synthetic approach used in this book is the one most commonly utilized by chemists who are professionally engaged in organic synthesis.

Like most books, this one contains several drawbacks. The book is wordy and in some cases redundant. However, most teachers know that one approach to a problem seldom gets an idea across to all students. Normally in class several approaches are tried in the hope that one will strike home. This book is founded on that premise.

The synthesis problems in the early chapters are ridiculously easy. No problem is easy to a person who cannot do it. In addition, this is the place where the ground work for the question-answer format, described above, is laid. The format of working problems backwards is difficult for most students to comprehend; therefore, the first exposure to this format should be as simple as possible.

The synthetic procedures are limited. I submit that this book is not intended as an organic text; it is designed to supplement an organic text with the express purpose of teaching students a method of organic synthesis. A complete listing of all the organic reactions would fill a large text and its cost would be prohibitive. I have endeavored to select the most frequently used synthetic procedures and in several places in the book, including a special instruction page, encouraged the student to frequently consult his text for other procedures to specific syntheses.

The organization of the book does not fit the format of all organic texts. My answer to that statement is that I have used this book successfully in various forms for the past five years while using the following texts:

Organic Chemistry by Morrison and Boyd
Modern Organic Chemistry by Roberts and Caserio
Organic Chemistry by Allinger, Cava, *et al.*

In summary, I would like to state that this book will not help all students taking organic chemistry. Many students seem to have an intuitive approach to organic synthesis problems. Other students will find this book very helpful. It is for these students that this book is written. With these students in mind, I feel that the advantages of this book far outweigh its disadvantages.

M. Paul Servé, July, 1975

ACKNOWLEDGMENT

I would like to thank Dr. William Feld and Mr. Thomas Mazer for their suggestions and encouragement of this endeavor and Miss Fern Davis for her patience in typing and retyping this manuscript.

TO THE STUDENT

This book is intended as a supplement to an organic chemistry text and is in no way meant as a substitute for the aforementioned text. The procedures outlined in this text do not and are not intended to cover all the synthetic routes available for the syntheses of organic molecules. Thus, the student is cautioned at the outset to frequently refer to his text for other synthetic procedures which may be more useful for designing specific molecules.

The synthetic approach used in this text is one which is used by most synthetic organic chemists. The approach is basically very simple. One begins the synthesis with the desired product and attempts to prepare the compound in one step from another compound A. If A is not a permitted starting material then it in turn is synthesized in one step from a molecule B. This technique is continued until one arrives at a permitted starting material. This technique of working the problem backwards from the desired product to a given starting material prevents the student from using a scatter-shot approach which he would be forced to use if he began with a starting material and attempted to reach a product. The student will also find that this approach will work for any type of synthetic problem.

This book is divided into chapters illustrating the preparations of functional groups. Each chapter is divided into a worked and unworked syntheses section. In the worked syntheses section, problems are completely worked out using the procedures given at the beginning of the chapter. The student, when he develops a sufficient degree of sophistication in the area of organic synthesis, is encouraged to devise other synthetic routes to the same molecule. If the student finds that he understands all the worked syntheses, he should go to the unworked syntheses section and do the problems listed. The student may then refer to his organic text for further problems. For advanced problems the student is encouraged to use the index which is arranged according to functional group preparation.

As he progresses the student will discover that organic synthesis is an art. Constant practice will lead to a greater proficiency in the area of organic synthesis and at the same time, a deeper understanding of all the areas of organic chemistry.

TABLE OF CONTENTS

Chapter

This chapter will be concerned with the syntheses of both acyclic and cyclic alkanes.

PROCEDURES FOR SYNTHESIZING ALKANES

I-1. Reduction of an alkyl halide (RX)

$$X = F, Cl, Br, I$$

(a) Use of a Grignard Reagent

$$RX + Mg \rightarrow RMgX \xrightarrow{H_2O} RH$$
$$\xrightarrow{D_2O} RD$$

$$CH_3CH_2CH_2Br \xrightarrow{Mg} CH_3CH_2CH_2MgBr \xrightarrow{D_2O} CH_3CH_2CH_2\text{-}D$$

(b) Use of lithium aluminum hydride (LiAlH$_4$)

$$RX \xrightarrow{LiAlH_4} RH$$

$$RX \xrightarrow{LiAlD_4} RD$$

$$CH_3CH_2CH_2\underset{\underset{I}{|}}{C}HCH_3 \xrightarrow{LiAlH_4} CH_3CH_2CH_2\overset{\overset{H}{|}}{C}HCH_3$$

(c) Use of lithium borohydride ($LiBH_4$)-lithium hydride (LiH) mixture. (This reagent is specific for RX).

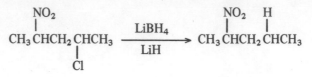

I-2. Reduction of an olefin ($\diagup C=C \diagdown$) or an alkyne (-C≡C-). This reaction requires the use of a catalyst. The commonly used catalysts are Ni, Pd, and Pt.

(a)

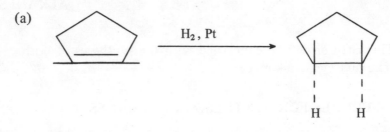

Get *cis*-addition of H, *i.e.*, both H are attached to the same side of $\diagup C=C \diagdown$.

(b) To place two D on a molecule

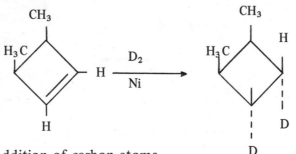

I-3. Addition of carbon atoms.

(a) Increasing the carbon chain length; Corey and House modification of the Wurtz reaction

$$RX \xrightarrow{\text{Li}} R\text{-Li} \xrightarrow{\text{CuX}} R_2CuLi \xrightarrow{R'X} R\text{-}R'$$

Note: For good yields R'X should be a primary halide, where RX may be a primary, secondary or tertiary halide.

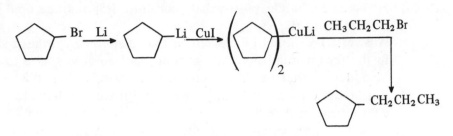

(b) Formation of cyclopropane rings. This reaction requires the use of diazomethane ($CH_2 N_2$) and ultraviolet light (hν).

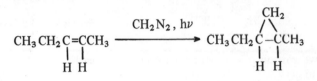

WORKED SYNTHESES

Using any alkyl halide or olefin containing no more than four carbon atoms, any monosubstituted cyclic molecule, and any inorganic reagents, prepare the following compounds: (Note: we will consider the following functional groups as substituents: D, F, Cl, Br, I, $\diagup$C=C$\diagdown$, -C≡C-).

I-a.
$$CH_3 CH_2 \overset{\overset{\displaystyle CH_3}{|}}{C}HCH_3$$

Let us ask some questions concerning I-a.

(1) What kind of a molecule is I-a?
    Answer: I-a is an acyclic alkane (2-methylbutane).

(2) Do we have to add carbon atoms?
    Answer: Since I-a contains five carbon atoms and we are permitted to use molecules containing no more than four carbon atoms, we must add carbon atoms.

(3) What procedures are available for adding carbon atoms?
    Answer: Procedures I-3a and I-3b. I-3b is used for the preparation of cyclopropane rings. Thus, I-3a is the procedure to use.

(4) How can I-a be cleaved so that procedure I-3a may be used
to synthesize it?
Answer: I-a should be cleaved so that it may be synthesized
in the fewest number of steps. Since carbon atoms have to
be added so that the total number of carbons is five, this
can be done by adding one molecule with four carbons to
another molecule with one carbon or by adding one molecule with three carbons to another molecule with two carbons. The breaking up of I-a is merely the reverse of the
above.

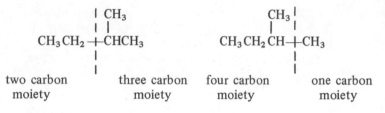

| two carbon | three carbon | four carbon | one carbon |
| moiety | moiety | moiety | moiety |

Let us now synthesize the molecule I-a using procedure I-3a
whereby we will add a 3-carbon molecule to a 2-carbon
molecule. Remember for procedure I-3a the stipulation
that one of the alkyl halides added must be a primary
halide. Looking at the two carbons which are to be joined
we can readily see that $CH_3 CH_2-$ is a primary carbon,
while $(CH_3)_2 CH-$ is a secondary carbon. This means that
for the synthesis of I-a, the isopropyl group must be in the
form of the CuLi complex. Thus we need compound **A**
and $CH_3 CH_2 I$ (a permitted starting material).

$$\underset{(CH_3 \overset{\displaystyle |}{C}H)_2 -CuLi}{CH_3} \qquad\qquad \textbf{A}$$

(5) How can **A** be prepared in one step?
Answer: By treatment of **B** with CuI as shown in procedure
I-3a.

$$\underset{CH_3 \overset{\displaystyle |}{C}H-Li}{CH_3} \qquad\qquad \textbf{B}$$

(6) How can **B** be prepared?
Answer: By treating **C** with Li as shown in procedure I-3a.

$$CH_3CH\text{-}I \overset{CH_3}{|} \qquad\qquad \mathbf{C}$$

Note: **C** is a permitted starting material.

Starting with the desired product and working backwards in a stepwise fashion, the synthesis of I-a becomes

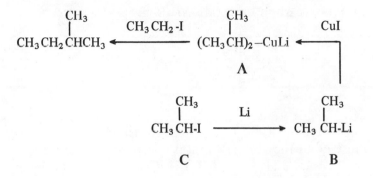

It should be noted that in each step the question, "How can this material be prepared?" is asked. This focuses our attention on the preparation of essentially a new product and requires us only to synthesize this newly created product in our synthetic sequence via a one-step reaction.

I-b

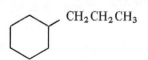

Let us ask ourselves some questions concerning I-b.

(1) What kind of a molecule is I-b?
Answer: I-b is a monosubstituted cycloalkane (propylcyclohexane).

(2) Do we have to add carbon atoms?
Answer: Yes, because we are permitted to use only cycloalkenes and cycloalkyl halides.

(3) What procedures are available for adding carbon atoms?
Answer: Procedure I-3a is the only procedure available for adding carbon atoms and not forming a cyclopropane molecule.

(4) How can I-b be broken up so that the required starting materials for synthesizing it via procedure I-3a can be discerned?
Answer: In the preparation of most ring compounds, the bond which should be cleaved is the bond from a ring carbon-to-carbon on a side chain.

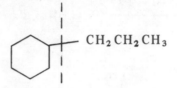

The $CH_3 CH_2 CH_2 -$ is a primary carbon. Thus the ring must be in the copper lithium complex. This means we need compound **A** and $CH_3 CH_2 CH_2 I$ (a permitted starting material).

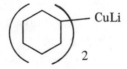   **A**

(5) How can **A** be prepared?
Answer: By treating **B** with CuI via procedure I-3a.

                                        B

(6) How can **B** be prepared?
Answer: By treatment of **C** with Li as shown in procedure I-3a.

                                        C

C is a permitted starting material.

Let us now synthesize I-b starting with the desired product and working backwards in a stepwise fashion until we arrive at a permitted starting material.

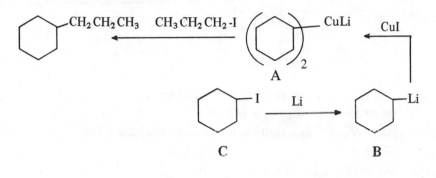

I-c

$$CH_3 \overset{\overset{\displaystyle CH_2}{\diagdown \diagup}}{CH}\text{-}CHCH_3$$

Let us ask some questions concerning I-c.

(1) What kind of molecule is I-c?
Answer: I-c is a disubstituted cyclopropane (1,2-dimethyl-cyclopropane).

(2) Do we have to add carbon atoms?
Answer: Yes, since we are not permitted to start with alkyl substituted cycloalkanes.

(3) What procedure is available for preparing cyclopropanes?
Answer: Procedure I-3b which requires the use of $CH_2 N_2$ and an olefin.

(4) What olefin is required?
Answer: Any carbon of the cyclopropane ring which possesses two H may be supplied by $CH_2 N_2$. In the case of I-c there is only one ring carbon that possesses two H. Thus, the other two carbons of the cyclopropane ring must be the olefin carbons.

Starting once again with the desired product and working backwards

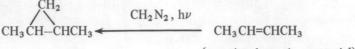

$$CH_3\overset{\overset{\displaystyle CH_2}{\diagup\diagdown}}{CH-CH}CH_3 \xleftarrow{\quad CH_2N_2,\ h\nu \quad} CH_3CH=CHCH_3$$

(permitted starting material)

I-d

$$CH_3CH_2\overset{\overset{\displaystyle CH_3}{|}}{CH}CH_2CH_3$$

Let us ask some questions concerning I-d.

(1)   What kind of molecule is I-d?
Answer: I-d is a non-cyclic (acyclic) alkane (3-methyl-pentane).

(2)   Do we have to add carbon atoms?
Answer: Yes, since I-d contains six carbons and we are permitted to use nothing larger than a 4-carbon acyclic molecule.

(3)   What procedure is available for adding carbon atoms to molecules?
Answer: I-3a is the only procedure available for the pre-paration of non-cyclopropane molecules.

(4)   How can I-d be cleaved so that it may be synthesized using procedure I-3a?
Answer: The two ways available of forming a 6-carbon molecule starting with a molecule containing no more than four carbons are either adding a 3-carbon molecule to a 3-carbon molecule, or adding a 2-carbon molecule to a 4-carbon molecule. Looking at I-d there is no bond which can be cleaved to give two 3-carbon moieties. However, there are two bonds which can be cleaved that will result in a 2-carbon and a 4-carbon moiety.

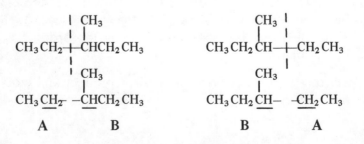

The carbon underlined in **A** is a primary carbon, while the carbon underlined in **B** is a secondary carbon. From I-3a it is readily apparent that **B** must be attached to the copper and lithium. Thus we need compound **C** and ethyl iodide.

$$\underset{\text{(CH}_3\text{CH}_2\text{CH)}_2\text{—CuLi}}{\overset{\overset{\text{CH}_3}{|}}{}}$$

C

(5) How can **C** be prepared?
Answer: By treating **D** with CuI.

$$\underset{\text{CH}_3\text{CH}_2\text{CH-Li}}{\overset{\overset{\text{CH}_3}{|}}{}}$$

D

(6) How can **D** be prepared?
By reacting **E** with Li

$$\underset{\text{CH}_3\text{CH}_2\text{CH-I}}{\overset{\overset{\text{CH}_3}{|}}{}}$$

E

E is a permitted starting material.

Let us now synthesize I-d starting with the desired product and working backwards in a stepwise manner until we arrive at a permitted starting material.

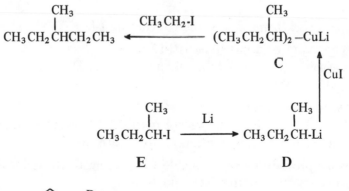

I-e

Let us ask some questions concerning I-e.

(1) What kind of a molecule is I-e?
Answer: I-e is a cycloalkane which has two D atoms on vicinal carbons (1,2-dideuterocyclopentane).

(2) Do we have to add carbon atoms?
Answer: No, since we can start with cycloalkyl halides and cycloalkenes.

(3) What procedure is available for placing two deuterium atoms in a vicinal relationship on a molecule?
Answer: Procedure I-2b, which requires an olefin.

(4) What is the structure of the olefin required?
Answer: The double bond in the olefin must be between the carbon atoms which contain the D in the product.

Let us now synthesize I-e starting with the desired product and working backwards in a stepwise fashion until we come to a permitted starting material.

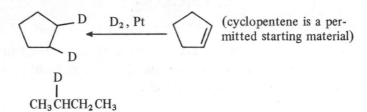

(cyclopentene is a permitted starting material)

I-f

$$CH_3\overset{\overset{\textstyle D}{|}}{C}HCH_2CH_3$$

Let us ask some questions concerning I-f.

(1) What kind of a molecule is I-f?
Answer: It is an acyclic alkane with one D atom (2-deuterobutane).

(2) Do we have to add carbon atoms?
Answer: No, because we are permitted to start with alkyl halides and olefins containing four carbons.

(3) What procedures are available for placing D on molecules?
Answer: Procedures I-1a and I-1b.

(4) Using procedure I-1a what is the structure of the Grignard reagent needed, which upon reaction with D_2O, will yield I-f?

Answer: Looking at procedure I-a it can be seen that the carbon containing the deuterium in the product must contain the MgBr in the starting material; thus the structure of the Grignard reagent needed is **A**.

$$CH_3CHCH_2CH_3$$
$$|$$
$$MgBr$$

A

(5) If procedure I-1b is used, what is the structure of the alkyl halide required, which upon reaction with $LiAlD_4$ will yield I-f?

Answer: In this case the carbon which contains the deuterium in the product must contain a halogen in the starting material. If Br is used as the halogen, we need **B**.

$$CH_3CHCH_2CH_3$$
$$|$$
$$Br$$

B

Let us now synthesize I-f using procedure I-1a. We will once again start with the desired product and work backwards in a stepwise manner.

$$CH_3CHCH_2CH_3 \xleftarrow{\quad D_2O \quad} CH_3CHCH_2CH_3 \xleftarrow{\quad Mg \quad} CH_3CHCH_2CH_3$$
$$\underset{D}{|} \qquad\qquad \underset{MgBr}{|} \qquad\qquad \underset{Br}{|}$$
$$\qquad\qquad\qquad\qquad\qquad \textbf{A} \qquad\qquad\qquad\qquad \textbf{B}$$

If procedure I-1b is used, the synthesis becomes

$$\underset{CH_3CHCH_2CH_3}{\overset{D}{|}} \xleftarrow{\quad LiAlD_4 \quad} \underset{CH_3CHCH_2CH_3}{\overset{Br}{|}}$$
$$\textbf{B}$$

I-g

Let us ask some questions concerning I-g.

(1) What kind of molecule is I-g?
Answer: I-g is an alkane made up of a six-membered ring fused to a three-membered ring (bicyclo[4.1.0]heptane).

(2) Do we have to add carbon atoms?
Answer: Yes, because we are permitted to start with only a cycloalkyl halide or a cycloolefin.

(3) What procedures are available for adding carbon atoms?
Answer: Procedures I-3a and I-3b. Since I-g can be looked upon as possessing a cyclopropane ring, let us use procedure I-3b.

(4) What is the structure of the olefin required so that I-g can be prepared by the use of CH_2N_2?
Answer: The CH_2 of the cyclopropane ring is supplied by the CH_2N_2. The other carbons making up the cyclopropane ring must contain the double bond.

Starting with the desired product I-g and working backwards in a stepwise fashion, the synthesis becomes simply the following

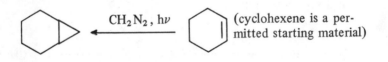

CH_2N_2, hν (cyclohexene is a permitted starting material)

UNWORKED SYNTHESES

Synthesize the following using the instructions set forth at the beginning of the worked syntheses section.

(h) $CH_3CH_2CH_2CH_2CH_2CH_2CH_3$

(i) cyclohexane–CH_2CHCH_3 with CH_3 substituent

(j) $CH_3CHCH_2CHCH_3$ with CH_3 and CH_3 substituents

(k)

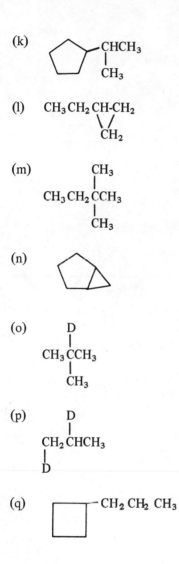

(l) $CH_3 CH_2 CH\text{-}CH_2$
                $\diagdown / \diagup$
                    CH_2

(m) CH_3
                    $|$
      $CH_3 CH_2 \overset{}{C}CH_3$
                    $|$
                  CH_3

(n)

(o) D
             $|$
    $CH_3 \overset{}{C}CH_3$
             $|$
           CH_3

(p) D
             $|$
    $CH_2 \overset{}{C}HCH_3$
      $|$
      D

(q) $-CH_2 CH_2 CH_3$

CHAPTER II

ALKENES (OLEFINS)

PROCEDURES FOR SYNTHESIZING ALKENES

There are three procedures which are frequently used in the syntheses of alkcncs.

II-1.* Dehydrohalogenation of an alkyl halide (R-X) X = F, Cl, Br and I.

$$
\underset{\overset{|}{Br}}{\overset{\overset{CH_3}{|}}{CH_3 CHCHCH_3}} \xrightarrow{KOH, \Delta} \underset{(70\%)}{\overset{\overset{CH_3}{|}}{CH_3 C=CHCH_3}} + \underset{(30\%)}{\overset{\overset{CH_3}{|}}{CH_3 CHCH=CH_2}}
$$

In this reaction, the halogen and a H from an adjacent carbon are removed.

II-2* Dehydration of an alcohol (R-OH).

$$
\underset{\overset{|}{OH}}{\overset{\overset{H}{|}}{CH_3 CHCCH_3}} \xrightarrow{acid, \Delta} \underset{(80\%)}{CH_3 CH=CHCH_3} + \underset{(20\%)}{H_2 C=CHCH_2 CH_3}
$$

The acids normally used are $H_2 SO_4$, $H_3 PO_4$ and $Al_2 O_3$.

_____

*Procedures II-1 and II-2 each yield a mixture of olefin products. However, the olefin formed as the major product will be the one which has the most alkyl groups attached to the olefinic carbons (more highly substituted olefin).

15

In this reaction the OH and a H from an adjacent carbon are removed.

II-3.   Reduction of an alkyne (R-C≡C-R).

  (a) Catalytic method
     Catalysts used (1) Lindlars Catalysts Pd/C
                   (2) Ni-B (P-2)

Both of the above catalysts yield a *cis*-addition product (H- adds to the same side of the C≡C).

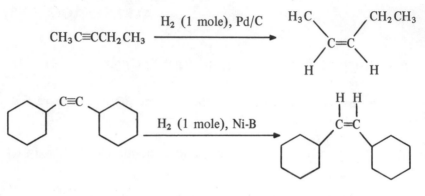

  (b) Chemical method
     Na and NH_3 (liquid) are the required reagents. The product is the *trans*-olefin.

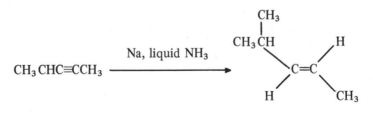

WORKED SYNTHESES

For problems II-a through II-c select a properly substituted alkyl halide or alcohol so that the desired alkene will be the major product of the elimination reaction. (Since there is no limitation as to the number of carbon atoms in the starting material, we do not have to worry about reactions which add carbons. Nevertheless, we shall continue the same question-answer format established in the alkane section.)

II-a

$$CH_3CH_2\overset{\overset{\displaystyle CH_2}{\displaystyle \|}}{C}CH_3$$

Let us ask some questions concerning II-a.

(1) What kind of molecule is II-a?
Answer: II-a is a terminal acyclic olefin (2-methyl-l-butene).

(2) What procedures are available for the preparation of acyclic olefins?
Answer: Procedures II-1, II-2 and II-3. Let us use procedure II-1, the dehydrohalogenation of an alkyl halide.

(3) What is the structure of the alkyl halide that will undergo dehydrohalogenation and yield II-a as the major product?
Answer: The olefin is formed by removing the elements of HX from two adjacent carbon atoms (olefinic carbon atoms in product). Thus, we merely have to add the elements of HX to the desired product to find the structure of the required alkyl halide. This can be accomplished in the two ways **A** and **B** shown below. We shall use Br as the halogen, although Cl and I will also work.

A. Adding Br to C_1 and H to C_2

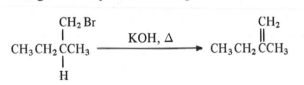

B. Adding H to C_1 and Br to C_2

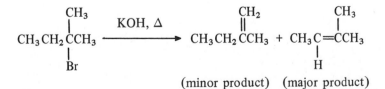

(minor product)   (major product)

As is evident adding the Br according to **A** will yield II-a as the only product upon reaction with KOH. However, adding

the Br according to B will yield II-a as the minor product
upon reaction with KOH. Thus, the structure of the alkyl
bromide required is

$$CH_3CH_2CCH_3$$

with CH_2Br above and H below the central carbon.

II-b   CH₃

Let us ask some questions concerning II-b.

(1) What kind of molecule is II-b?
    Answer: II-b is a cycloalkene (l-methylcyclopentene).

(2) What procedures are available for preparing a cycloalkene?
    Answer: Procedures II-1 and II-2. Let us select II-2, the
    dehydration of an alcohol.

(3) What is the structure of the alcohol that through the use of
    procedure II-2 will yield II-b as the major product?
    Answer: The dehydration of an alcohol involves the removal
    of the elements of water from two adjacent carbon atoms
    (olefinic carbons in product). Thus, the addition of the
    elements of water to the olefinic carbons will give us the
    structure of the required alcohol. The addition of the H
    and OH may be accomplished in the two ways **A** and **B**
    shown below.

A. Adding OH to C_1 and H to C_2

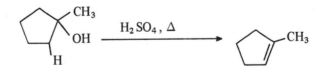

B. Adding H to C_1 and OH to C_2

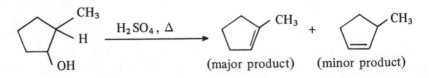

Both alcohols would yield the desired olefin upon dehydration. However, path **A** yields only the desired product while path **B** yields two products. Thus the required alcohol is

II-c

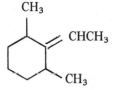

Let us ask some questions concerning II-c.

(1) What kind of a molecule is II-c?
Answer: II-c is a trisubstituted olefin (2,6-dimethylethyl-idenecyclohexane).

(2) What procedures are available for the preparation of trisubstituted olefins?
Answer: Procedures II-1, II-2 and II-3. Let us use procedure II-1, the dehydrohalogenation of an alkyl halide.

(3) What is the structure of the alkyl halide required to produce II-c by procedure II-1?
Answer: Let us assume that the halogen is chlorine. The olefinic bond is formed by removing a Cl and a H from two adjacent carbon atoms (olefinic carbons in product). To reconstruct the alkyl chloride required, we merely add a Cl and a H to the olefinic carbons of II-c. This can be done in two ways **A** and **B**.

A. Adding H to the ring olefinic carbon and Cl to the non-ring olefinic carbon:

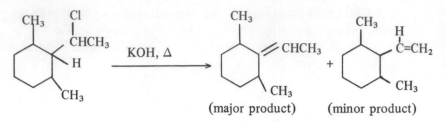

(major product)   (minor product)

B. Adding Cl to the ring olefinic carbon and H to the non-ring olefinic carbon:

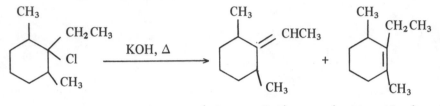

(minor product)   (major product)

As can easily be seen, the structure of the alkyl halide which will yield II-c as the major product of the dehydrohalogenation reaction is

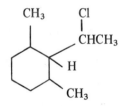

For problems II-d and II-e use a properly substituted alkyne. (Note stereochemistry)

II-d

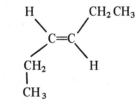

Let us ask some questions about II-d.

(1) What kind of a molecule is II-d?
Answer: II-d is a dialkyl *trans*-olefin (*trans*-3-hexene).

(2) What procedure is available whereby starting with an alkyne a *trans*-olefin results upon reduction?
Answer: Procedure II-3b.

(3) Where must be the -C≡C- be located?
Answer: The olefin carbon atoms in the product are the triple bonded carbon atoms in the starting alkyne. Working backwards from the desired product the synthesis thus becomes

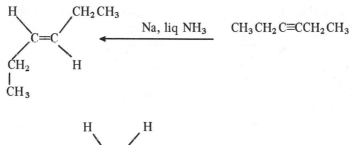

II-e

Let us ask some questions concerning II-e.

(1) What kind of a molecule is **II-e**?
Answer: II-e is a *cis*-disubstituted olefin

cis-1-(2-methylcyclopentyl)-1-propene

(2) What procedure is available for preparing *cis*-olefins by the reduction of an alkyne?
Answer: Procedure II-3a. In this case the use of either the Pd/C catalyst or the Ni-B (P-2) catalyst will reduce the alkyne carbon atoms to olefin carbons with the required *cis*-stereochemistry.

Starting with the desired product and working backwards stepwise the synthesis of II-e becomes

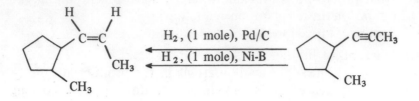

UNWORKED SYNTHESES

Using the same conditions as stipulated for the syntheses of II-a through II-c prepare the following:

(f)

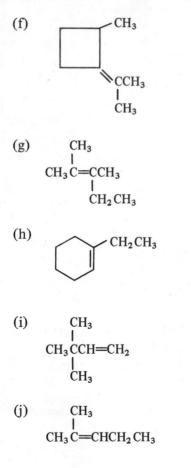

(g)

$$CH_3 \overset{\underset{\displaystyle CH_3}{|}}{\underset{\underset{\displaystyle CH_2CH_3}{|}}{C}} = CCH_3$$

(h)

(i)

$$CH_3 \overset{\underset{\displaystyle CH_3}{|}}{\underset{\underset{\displaystyle CH_3}{|}}{C}} CH = CH_2$$

(j)

$$CH_3 \overset{\underset{\displaystyle CH_3}{|}}{C} = CHCH_2CH_3$$

Synthesize the following from the properly substituted acetylene.

(k)

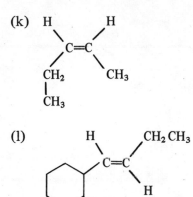

(l)

H   CH₂CH₃
 \ /
  C=C
 /   \
      H
CH₂CH₃

PROCEDURES FOR SYNTHESIZING ALKYL HALIDES

Presented below are some of the most frequently used methods of preparing alkyl halides. It is important to note that some of the reactions permit the incorporation of one halogen into a molecule, while other reactions allow polyhalogenation of a molecule.

III-1 Addition of HX to an olefin (X = Cl, Br, I).

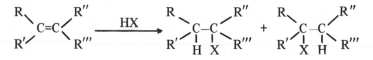

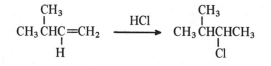

This reaction follows the Markovnikov addition rule which states that the H or (+) portion of the addendum adds to the olefinic carbon having the most hydrogens.

III-2 Addition of HBr and peroxide (R-O-O-R) to an olefin.

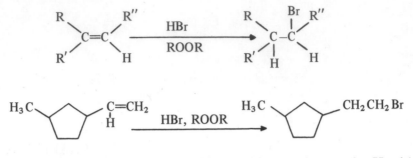

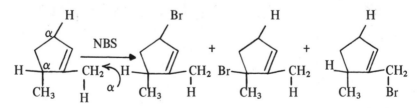

This reaction follows anti-Markovnikov addition, the H adds to the olefinic carbon containing the fewest hydrogens.

III-3 N-Bromosuccinimide (NBS) reaction

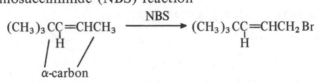

This reaction is unique because it permits one to place a bromine atom on a molecule containing an olefin linkage and the product retains the C=C. For this reaction to take place there must always be a H on an α— or allylic carbon (carbon adjacent to C=C). This is the H that is replaced by the Br. If there are several α-carbons each containing H, then the Br may go to any of the α-carbons, *i.e.,*

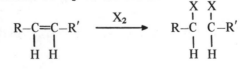

The number of potential products and the ease with which they can be separated determines the utility of this reaction.

III-4 Addition of X_2 to an olefin.

$$\underset{\underset{H\ H}{|\ \ |}}{R-C=C-R'} \xrightarrow{X_2} \underset{\underset{H\ H}{|\ \ |}}{\overset{\overset{X\ X}{|\ \ |}}{R-C\ C-R'}}$$

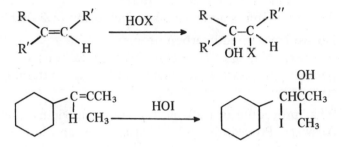

The product is a dihalide in which the halogens are arranged in a vicinal (on adjacent carbons) manner.

III-5 Addition of HOX to an olefin.

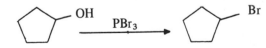

The product is a molecule containing a halogen and a hydroxyl group (OH) arranged in a vicinal orientation. The reaction follows Markovnikov addition with the halogen being the (+) portion of the addendum. HOX may also be written as $H_2O + X_2$. Therefore, HOCl may also be written as $H_2O + Cl_2$ in a reaction sequence.

III-6 Reaction of an alcohol with PX_3 (X = Cl and Br)

$$ROH + PX_3 \longrightarrow R\text{-}X$$

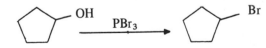

$$CH_3 CH_2 CH_2\text{-}OH \xrightarrow{PCl_3} CH_3 CH_2 CH_2\text{-}Cl$$

WORKED SYNTHESES

Using any monosubstituted acyclic or cyclic organic molecule containing no more than six carbons and any inorganic reagents, prepare the following compounds. (Note: monosubstituted means the molecule may contain an olefinic linkage, or a halogen, but not both.)

III-a

$$\underset{\underset{Cl}{|}\quad\underset{Cl}{|}}{CH_2\ CHCH_2\ CH_3}$$

Let us ask some questions concerning this molecule.

(1) What kind of a molecule is III-a?
Answer: III-a is a vicinal alkyl dichloride (1,2-dichlorobutane).

(2) Do we have to add carbon atoms?
Answer: No, since III-a contains four carbons and we are permitted to start with organic molecules containing up to six carbons.

(3) What procedure is available for preparing vicinal dichlorides?
Answer: Procedure III-4 which requires an olefin.

(4) What is the structure of the required olefin?
Answer: The carbons which contain the Cl atoms in the product must be the olefinic carbons in the desired olefin. Thus the olefin needed is $H_2\,C{=}CHCH_2\,CH_3$, which is a permitted starting material.

Thus, starting with the desired product and working backwards in a stepwise manner, the synthesis becomes

$$\underset{\underset{Cl}{|}\quad\underset{Cl}{|}}{CH_2\ CHCH_2\ CH_3} \xleftarrow{\quad Cl_2 \quad} H_2C{=}CHCH_2\ CH_3$$

III-b

$$\underset{}{\overset{\overset{\textstyle CH_3}{|}}{Br\text{-}CH_2\ C{=}CHCH_3}}$$

Let us ask some questions concerning III-b.

(1) What kind of molecule is III-b?
Answer: III-b is an alkene which has a Br located on the α-(allylic) carbon (1-bromo-2-methyl-2-butene).

(2) Do we have to add carbon atoms?
Answer: No, since we are permitted to start with molecules containing up to six carbons and since III-b contains only five carbons.

(3) What procedure is available for placing a Br on an α-carbon atom?
Answer: Procedure III-3, which required NBS and an olefin.

(4) What is the structure of the required olefin?
Answer: The structure of the olefin would have the same carbon skeleton as III-b. The only difference would be that a H would replace the Br in III-b. In other words, the double bond is still between C_2 and C_3 in the required olefin. The required olefin is

$$\overset{\displaystyle \overset{CH_3}{|}}{H\text{-}CH_2C}\text{=}CHCH_3$$

which is a permitted starting material.

Let us now synthesize III-b starting with the desired product and working backwards in a stepwise fashion.

$$\overset{\displaystyle \overset{CH_3}{|}}{Br\text{--}CH_2C}\text{=}CHCH_3 \xleftarrow{\quad NBS \quad} \overset{\displaystyle \overset{CH_3}{|}}{CH_3C}\text{=}CHCH_3$$

It should be noted that NBS on the above olefin will also yield

$$\overset{\displaystyle \overset{CH_3}{|}}{CH_3C}\text{=}\underset{\displaystyle \underset{Br}{|}}{CHCH_2}$$

as a product. However, looking at the starting olefin above, one can easily see that since there are six α-H attached to the olefinic carbon C_2 while olefinic carbon C_3 has only three α-H attached; statistically the bromination will favor the production of the desired product III-b by a factor of 2:1.

III-c

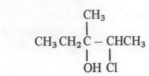

$$CH_3CH_2\overset{\overset{\displaystyle CH_3}{|}}{C}-\underset{\underset{\displaystyle Cl}{|}}{\overset{\overset{\displaystyle }{|}}{C}}HCH_3$$

OH

Let us ask some questions concerning III-c.

(1) What kind of a molecule is III-c?
Answer: III-c is a vicinal chloroalcohol (2-chloro-3-methyl-3-pentanol).

(2) Do we have to add carbon atoms?
Answer: No, because III-c contains only six carbons and we are permitted to start with organic molecules containing six carbons.

(3) Is there a procedure available for preparing vicinal chloro-alcohols in one step?
Answer: Procedure III-5, which requires HOCl and an olefin.

(4) What is the structure of the olefin required?
Answer: The Cl and OH add across the olefinic linkage. Thus, the carbons that contain the Cl and OH in the product must be the olefinic carbons in the required olefin. The required olefin would be

$$CH_3CH_2\overset{\overset{\displaystyle CH_3}{|}}{C}=CHCH_3$$

which is a permitted starting material.

Working the problem backwards in a stepwise fashion, the synthesis of III-c becomes

$$CH_3CH_2\overset{\overset{\displaystyle CH_3}{|}}{\underset{\underset{\displaystyle OH\ Cl}{|\ \ \ |}}{C}}-CHCH_3 \quad\xleftarrow[\text{(H}_2\text{O + Cl}_2)]{\text{HOCl}}\quad CH_3CH_2\overset{\overset{\displaystyle CH_3}{|}}{C}=CHCH_3$$

III-d

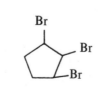

Let us ask some questions concerning III-d.

(1)   What kind of a molecule is III-d?
       Answer: III-d is a tribrominated cyclopentane molecule in
       which the Br are on adjacent carbon atoms (1,2,3-tribromo-
       cyclopentane).

(2)   Do we have to add carbon atoms?
       Answer: No, because III-d contains only five carbons and
       we are permitted to start with organic molecules containing
       up to six carbons.

(3)   Is there a procedure available which will permit the placing
       of three Br in adjacent carbons in one step?
       Answer: No.

(4)   Is there a procedure available which would permit the place-
       ment of two Br atoms on a molecule in a vicinal orientation?
       Answer: Procedure III-4 which requires Br_2 and an olefin.

(5)   What would be the structure of the olefin required so that
       upon the addition of Br_2 via procedure III-4, III-d would be
       the product isolated?
       Answer: In order that III-4 be the product, one Br would
       have to already be in the molecule. Thus, the structure of
       the required olefin would be **A**.

(6)   Is **A** a permitted starting material?
       Answer: No, because it contains two functional groups
       (C=C and Br), and we are permitted to start with monosub-
       stituted organic molecules.

We are now faced with the synthesis of **A**. Let us ask some
questions concerning **A**.

(7)   What kind of a molecule is **A**?
       Answer: **A** is a cyclic olefin containing an ollylic Br.

(8)   Is there a procedure available for the preparation of allylic
       bromides?

Answer: Procedure III-3, which requires NBS and an olefin.

(9)   What is the structure of the required olefin?
Answer: Since NBS does not change the position of the double bond, the olefinic carbons in the product are the olefinic carbons in the starting material. Thus the required olefin would be

 B

B is a permitted starting material.

Let us now synthesize III-d starting with the desired product and working backwards in a stepwise manner until we come to a permitted starting material.

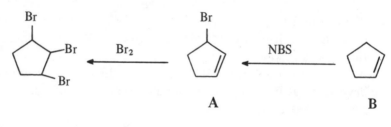

III-e

$$CH_2\ CH_2\ \underset{\underset{Cl}{|}}{\overset{\overset{CH_3}{|}}{C}}CH_3$$
| |
Br

Let us ask some questions concerning III-e.

(1)   What kind of molecule is III-e?
Answer: III-e is an alkyl dihalide (1-bromo-3-chloro-3-methylbutane).

(2)   How may a Cl be placed on an organic molecule?
Answer: Procedure III-1 is the only method currently available for placing one Cl in a molecule; HCl and an olefin are required.

(3)   How may a Br be placed on an organic molecule?
Answer: Procedures III-1, III-2 and III-3 will all permit the placing of one Br on an organic molecule; an olefin is required.

(4) How do we start the synthesis of III-e?
Answer: Since there is only one way of putting a Cl on a molecule, while there are three ways which can be used to place a Br on a molecule, let us start the synthesis by working back from the desired product and adding in the last step, the substituent which we have the least flexibility in incorporating into a molecule, namely the Cl.

(5) What is the structure of the olefin that by adding HCl III-e will be the product?
Answer: There are two possible olefins **A** and **B** which upon the addition of HCl will yield III-e.

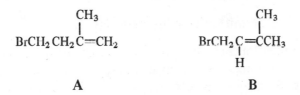

$$\underset{\textbf{A}}{\overset{\overset{\displaystyle CH_3}{|}}{BrCH_2\,CH_2\,C=CH_2}} \qquad\qquad \underset{\textbf{B}}{\overset{\overset{\displaystyle CH_3}{|}}{\underset{\underset{\displaystyle H}{|}}{BrCH_2\,C=CCH_3}}}$$

Thus the synthesis of III-e requires the preparation of either **A** or **B**.

(6) What kind of molecules are **A** and **B**?
Answer: **A** is a β-substituted bromoolefin (4-bromo-2-methyl-1-butene). **B** is an α-substituted bromoolefin (4-bromo-2-methyl-2-butene).

(7) Is there a procedure available for preparing a β-substituted bromoolefin in one step?
Answer: Yes, procedure III-1 which would require olefin **C**.

$$\underset{\underset{\displaystyle H}{|}}{\overset{\overset{\displaystyle CH_3}{|}}{H_2C=C-C=CH_2}} \qquad\qquad \textbf{C}$$

However in **C**, HBr can add across either olefinic linkage or it can even add to C_1 and C_4. Thus, the complications associated with this step are enormous. It should also be noted that **C** is not a permitted starting material and would have to be synthesized.

(8) Is there a procedure available for the preparation of α-substituted bromoolefins?

Answer: Yes, procedure III-3 requiring NBS and an olefin.

(9) What is the structure of the required olefin referred to in Answer 8?

Answer: Since NBS does not alter the position of the olefinic linkage, the carbons that are olefinic in the product all contain the double bond in the starting material. Thus the required olefin is

which is a permitted starting material.

A complication here is that NBS can easily brominate the other α-carbons to yield an undesired bromoolefin. Nevertheless, **B** can be separated from the side reaction products by distillation. Thus, the syntheses of **A** and **B** are not without complications. However, since **A** requires more steps and the chances for a good yield of the desired product decreases with the more steps involved, let us select **B** as the bromoolefin to be utilized in the synthesis of III-e.

The synthesis of III-e, working backwards in a stepwise manner, now becomes

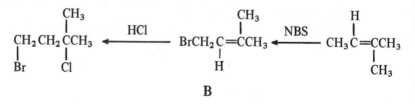

Note: When a synthesis requires the incorporation of more than one functional group into a molecule, it is best to work backwards from the desired product and in the last step, place on the molecule the substituent which have the fewest ways of incorporating into a molecule.

III-f

$$\underset{\underset{CH_3}{|}}{CH_3}\underset{}{CH}\underset{\underset{Cl}{|}}{CH}CHCH_2\text{-}Cl$$

Let us ask some questions about III-f.

(1) What kind of molecule is III-f?
Answer: III-f is an alkyl trihalide containing two Cl which are on vicinal carbon atoms (1,2-dichloro-3-bromo-4-methyl-pentane).

(2) Do we have to add carbon atoms?
Answer: No because we are allowed to start with organic molecules containing up to six carbons and III-f contains six carbons.

(3) Is there a procedure available for placing two vicinal Cl atoms on an organic molecule?
Answer: Yes, procedure III-4 which requires an olefin will suffice.

(4) What procedures will place a Br atom on an organic molecule?
Answer: Procedures III-1, III-2 and III-3 will work.

(5) How shall we commence the synthesis of III-f?
Answer: We shall work the synthesis back from the desired product. The last step will once again require placing on the molecule the substituent which we have the fewest ways of incorporating into the molecule, namely the Cl. Thus, olefin A is required and the synthesis now boils down to the preparation of A.

$$\underset{\underset{CH_3}{|}}{CH_3}CH\underset{\underset{H}{|}}{CH}C{=}CH_2 \qquad\qquad \textbf{A}$$

(6) What kind of molecule is A?
Answer: A is an acyclic olefin containing an allylic Br (3-bromo-4-methyl-1-pentene).

(7) Is there a procedure available for placing a Br on an α-carbon of an olefin?
Answer: Yes, procedure III-3 which will require olefin B, a permitted starting material.

$$CH_3\underset{\underset{CH_3}{|}}{C}HCH_2\underset{\underset{H}{|}}{C}=CH_2 \qquad B$$

Thus, the complete synthesis of III-f is

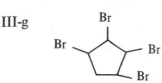

$$\underset{\underset{CH_3 \quad Cl}{|\quad |}}{CH_3CH\overset{\overset{Br}{|}}{C}HCHCH_2Cl} \xleftarrow{Cl_2} \underset{\underset{CH_3 \quad H}{|\quad |}}{CH_3CH\overset{\overset{Br}{|}}{C}HC=CH_2} \xleftarrow{NBS}$$

A

$$CH_3\underset{\underset{CH_3}{|}}{C}HCH_2\underset{\underset{H}{|}}{C}=CH_2$$

B

III-g

Let us ask some questions concerning III-g.

(1) What kind of molecule is III-g?
Answer: III-g is a tetrabromocyclopentane (1,2,3,4-tetra-bromocyclopentane).

(2) Do we have to add carbon atoms?
Answer: No, because III-g contains only five carbon atoms.

(3) Is there a procedure available for placing four Br in a vicinal arrangement on a molecule in one step?
Answer: Procedure III-4 permits the addition of two Br in one step across a double bond. If we had a molecule containing two olefinic linkages, we could add four Br in one step. The molecule required would be **A**.

(4) Is **A** a permitted starting material?
Answer: No, because it contains two C=C and thus is considered disubstituted.

(5) What kind of molecule is **A**?
Answer: **A** is a cyclic diolefin (1,3-cyclopentadiene).

(6) What procedure is available for the preparation of **A**?
Answer: Procedure II-1 will yield **A** upon treatment of either **B** or **C** with KOH.

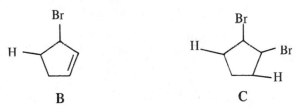

B   **C**

Treatment of **B** with one equivalent of KOH will yield **A** while treatment of **C** with two equivalents of KOH will yield **A**. Neither **B** nor **C** is a permitted starting material.

Let us use **C**. As an exercise the reader can use **B** as the precursor of **A**.

(7) What kind of molecule is **C**?
Answer: **C** is a vicinal dibromocyclopentane (1,2-dibromo-cyclopentane).

(8) Is there a procedure available for the preparation of vicinal dibrominated alkanes?
Answer: Yes, procedure III-4 which would require olefin **D**, a permitted starting material.

D

Thus the synthesis of III-g becomes

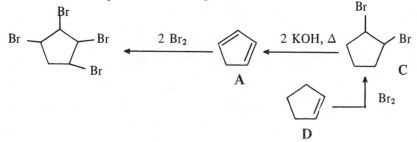

A

D

Another type of synthesis problem frequently encountered in organic chemistry concerns the synthesis of a molecule starting with a specified starting material. The next two syntheses illustrate that type problem. It should be kept in mind that the synthetic approach used previously will also work in these cases.

III-h Starting with

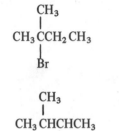

synthesize

$$\underset{\underset{\displaystyle Br}{|}}{\overset{\overset{\displaystyle CH_3}{|}}{CH_3\,CHCHCH_3}} \qquad \text{(III-h)}$$

(1) What kind of molecule is III-h?
Answer: III-h is a monobromoalkane (2-methyl-3-bromo-butane).

(2) What procedures are available for the preparation of monobromoalkanes?
Answer: Procedures III-1 and III-2 each of which requires an olefin.

(3) What are the structures of the required olefins whereby the use of procedures III-1 and III-2 would yield III-h?
Answer: III-1 adds HBr via a Markovnikov pathway, thus the olefin required would be **A**.

$$\underset{\underset{\displaystyle H}{|}}{\overset{\overset{\displaystyle CH_3}{|}}{CH_3\,CHC\!=\!CH_2}}$$

For procedure III-2 HBr is added via an anti-Markovnikov pathway, thus the required olefin is **B**.

 B

(4) What procedures are available for the preparation of olefin **A**?

Answer: **A** may be prepared by procedure II-1 (dehydrohalogenation of an alkyl halide) or procedure II-2 (dehydration of an alcohol). Since we are concerned with arriving at a starting material which is an alkyl bromide, we shall disregard procedure II-2.

(5) What is the structure of the alkyl bromide which would yield **A** through the use of procedure II-1?

Answer: The alkyl bromide would have to be substituted in such a manner that the olefin formed was the most highly substituted olefin. Thus, for **A** to be the product the structure of the alkyl bromide must be **C**.

$$CH_3CHCH_2CH_2Br \qquad \overset{CH_3}{|} \qquad C$$

(6) What procedures are available for the preparation of **B**?

Answer: The same procedures as in answer 4. We shall once again focus on procedure II-1 because we want to get back to an alkyl bromide as starting material.

(7) What are the structures of the alkyl bromides which would yield olefin **B** through the use of procedure II-1?

Answer: Once again **B** must be the most highly substituted olefin upon dehydrohalogenation. The alkyl bromides which would yield **B** as the major product are **D** and **E**.

Since **E** is the compound we wish to prepare, starting with it is absurd. However, **D** is the alkyl bromide we are requested to start with and gives the required olefin **B** upon treatment with KOH.

Thus, the complete synthesis is

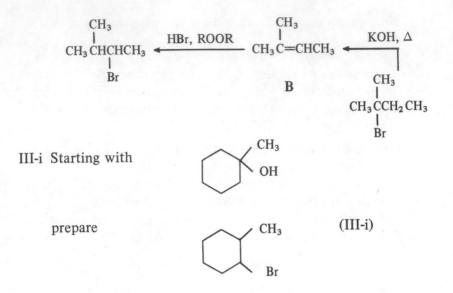

III-i Starting with

prepare   (III-i)

Let us ask some questions concerning III-i.

(1) What kind of molecule is III-i?
Answer: III-i is a cyclic alkyl bromide (1-bromo-2-methyl-cyclohexane).

(2) What procedures are available for the preparation of alkyl bromides?
Answer: Procedures III-1 and III-2 each of which requires an olefin.

(3) What are the structures of the required olefins so that via procedures III-1 and III-2, III-i would be the product?
Answer: Using procedure III-1 the only olefin which would yield III-i as the result of Markovnikov addition would be A.

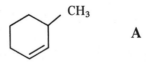

A

It should be noted that A would yield III-i only 50% of the time. HBr can just as easily add to A to give the other alkyl bromide.

Using procedure III-2, the olefin required would be B.

B

which would give III-i as the exclusive product.

In order to get the optimum yield from the synthetic route being proposed, we shall use **B** as the precursor of III-i.

(4) What procedures are available for the preparation of **B**?

Answer: **B** can be prepared by procedures II-1 (the dehydro-halogenation of an alkyl halide) and II-2 (the dehydration of an alcohol). Since we are restricted to starting with an alcohol, let us use procedure II-2.

(5) What is the structure of the alcohol which will yield **B** via procedure II-2?

Answer: Since the dehydration of an alcohol will yield the most highly substituted olefin and because the OH group must be attached to one of the carbon atoms which will become an olefinic carbon upon dehydration, alcohols **C** and **D** would each give **B** as the major product.

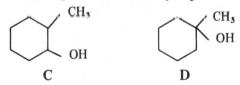

C   D

Note: **D** is the structure of the required starting material.

The entire synthetic route now becomes

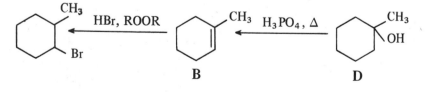

B   D

UNWORKED SYNTHESES

Prepare the following using the restrictions stipulated for the syntheses of III-a through III-g.

(j)

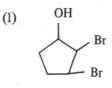

CH₃
|
CH₂CHCCH₂Cl
| | |
Cl Cl Cl

(k)

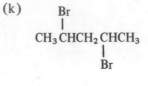

Br
|
CH₃CHCH₂CHCH₃
|
Br

(l)

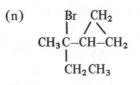

OH

Br

Br

(m)

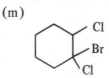

Cl

Br

Cl

(n)

Br CH₂
| / \
CH₃C–CH–CH₂
|
CH₂CH₃

(o)

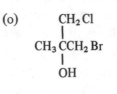

CH₂Cl
|
CH₃CCH₂Br
|
OH

(p)

CH₃
|
I–CH₂CCH₂CHCH₃
| |
OH   CH₃

(q)

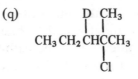

D CH₃
| |
CH₃CH₂CHCCH₃
|
Cl

Starting with

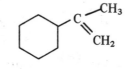

CH₃
C
CH₂

prepare the following

(r)

CH₃
|
CCH₃

(s)

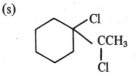

Cl

CCH₃

Cl

(t)

CH₂
\
CCH₃
|
CH₃

(u)

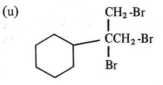

CH₂-Br
|
CCH₂-Br
|
Br

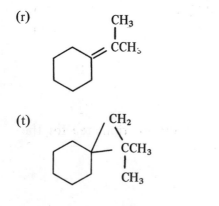

CHAPTER IV

ALKYNES (ACETYLENES)

Alkynes can be prepared through the use of two basic procedures. Before these procedures are presented, it should be noted
that the smallest ring in which an acetylene linkage is stable at
room temperature for an extended period of time is cyclooctyne.

PROCEDURES FOR SYNTHESIZING ALKYNES

IV-1   Removal of two moles of HX from an alkyl dihalide (X =
Cl, Br). The halogens may be vicinal (on adjacent carbon
atoms) or geminal (on the same carbon atom). If there
is a choice of more than one product being formed, the
more highly substituted alkyne will be formed in greater
abundance.

  (a)   Vicinal dehydrohalogenation

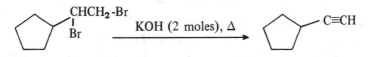

  (b)   Geminal dehydrohalogenation

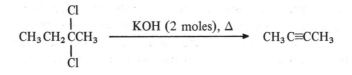

IV-2   Reaction of sodium acetylides with primary alkyl halides. (Excellent method for the addition of carbon atoms to molecules).

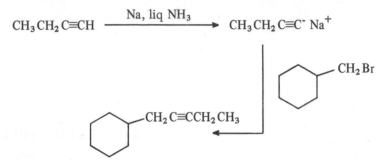

$$CH_3CH_2C\equiv CH \xrightarrow{\text{Na, liq } NH_3} CH_3CH_2C\equiv C^- Na^+$$

Note: For this reaction to proceed, there must be a H on an acetylenic carbon.

WORKED SYNTHESES

Using any monosubstituted cyclic molecule, any monosubstituted acyclic organic molecule containing no more than four carbons and any inorganic reagent, prepare the following:

IV-a CH_3
              |
        $CH_3CHCH_2C\equiv CCH_3$

Let us ask some questions concerning IV-a.

(1)   What kind of molecule is IV-a?
       Answer: IV-a is an acyclic alkyne (5-methyl-2-hexyne).

(2)   Do we have to add carbon atoms?
       Answer: Since the desired product contains seven carbons and we are allowed to use acyclic organic molecules containing no more than four carbons, yes, we must add carbon atoms.

(3)   What procedures are available for adding carbon atoms?
       Answer: Procedure I-3 (yields an alkane) and procedure IV-2 (yields an alkyne) are available. Since the desired product is an alkyne, we shall use procedure IV-2.

(4) How can IV-a be broken up into its component parts so that it may be resynthesized via procedure IV-2?
Answer: Since procedure IV-2 will form a carbon-carbon bond in which one of the carbons is an acetylenic carbon, then we must break one of the single bonds to an acetylenic carbon in IV-a. Thus for

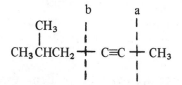

if we break the bond labelled (a) this will yield the two components **A** and $CH_3 I$ needed to resynthesize IV-a.

$$\underset{\displaystyle CH_3 CHCH_2 C\equiv C^- \ Na^+}{\overset{\displaystyle CH_3}{|}} \quad ICH_3 \qquad \textbf{A}$$

If we break the bond labelled (b), this would require **B** and **C**.

$$\underset{\displaystyle CH_3 CHCH_2\text{-Br}}{\overset{\displaystyle CH_3}{|}} \qquad\qquad Na^+ \ ^-C\equiv CCH_3$$

$$\textbf{B} \qquad\qquad\qquad \textbf{C}$$

A contains six carbons so **A** itself would have to be synthesized. **B** and **C** on the other hand contain four and three carbons respectively and are easily arrived at.

Let us now synthesize IV-a by breaking bond (b). Working backwards in a stepwise fashion, the synthesis of IV-a becomes

IV-b $HC\equiv CCH_2CH_2CH_3$

Let us ask some questions concerning IV-b.

(1)   What kind of molecule is IV-b?
      Answer: IV-b is an acyclic terminal acetylene (1-pentyne).

(2)   Do we have to add carbon atoms?
      Answer: Yes, since IV-b contains five carbons and we are permitted to use organic molecules containing no more than four carbons.

(3)   What procedures are available which permit the addition of carbon atoms?
      Answer: Procedure I-3 (yields an alkane) and procedure IV-2 (yields an alkyne) are available. Since IV-b is an alkyne, let us use procedure IV-2 as our method of adding carbon atoms.

(4)   How can IV-b be broken up into its component parts so that it may be resynthesized by procedure IV-2?
      Answer: Since procedure IV-2 forms a carbon-carbon bond between an alkyne and an adjacent carbon, the only bond available in IV-b for breaking would be the bond between C_2 and C_3. Thus IV-b can be broken down in the following way:

$$HC\equiv C\!\!+\!\!CH_2CH_2CH_3$$

$$HC\equiv C^-\ Na^+ \qquad Br\text{-}CH_2CH_2CH_3$$

$$\textbf{A} \qquad\qquad\qquad \textbf{B}$$

Note that both **A** and **B** contain less than four carbons. Therefore, they may be synthesized rather easily.

Working backwards in a stepwise manner, the synthesis of IV-b becomes

$$HC\equiv CCH_2CH_2CH_3 \xleftarrow{\quad CH_3CH_2CH_2\text{-}Br\quad} Na^+\ ^-C\equiv CH$$

$$\uparrow\ Na,\ liq\ NH_3$$

$$HC\equiv CH$$

IV-c   Prepare

$$CH_3CH_2C{\equiv}CCH_3$$

starting with

$$CH_3CH_2CH_2\underset{\underset{\displaystyle H}{|}}{C}{=}CH_2$$

Let us ask some questions concerning IV-c.

(1)   What kind of molecule is IV-c?
Answer: IV-c is a non-terminal alkyne (2-pentyne).

(2)   Do we have to add carbon atoms?
Answer: No, because the given starting material and IV-c each contain five carbons.

(3)   What procedure is available for preparing alkynes without adding carbon atoms?
Answer: Procedure IV-1, which requires a dihalide.

(4)   What would be the structure of the precursor which would yield IV-c via procedure IV-1?
Answer: Either the gem-dihalide **A** or the vicinal dihalide **B** would yield IV-c.

<table>
<tr><td>$$CH_3CH_2CH_2\underset{\underset{\displaystyle Br}{|}}{\overset{\overset{\displaystyle Br}{|}}{C}}CH_3$$</td><td>$$CH_3CH_2\underset{\underset{\displaystyle Br}{|}}{\overset{\overset{\displaystyle Br}{|}}{C}}HCHCH_3$$</td></tr>
<tr><td align="center">**A**</td><td align="center">**B**</td></tr>
</table>

We shall use the vicinal dihalide **B**. The reader may use **A** as the source of IV-c as an exercise.

(5)   How may **B** be prepared?
Answer: **B** may be prepared from either **C** or **D** via procedure III-4 or procedure III-1 as shown.

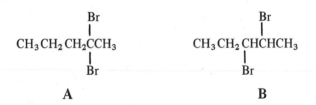

B $\xleftarrow{\quad Br_2 \quad}$ $CH_3CH_2\underset{\underset{\displaystyle }{}}{\overset{\overset{\displaystyle H}{|}}{C}}{=}CHCH_3$ **C**

B $\xleftarrow{\quad HBr \quad}$ $CH_3CH_2\underset{\underset{\displaystyle Br}{|}}{\overset{\overset{\displaystyle H}{|}}{C}}HC{=}CH_2$ **D**

(6) How may **C** be prepared?
Answer: **C** being an olefin may be prepared from **E** via procedure II-1.

$$CH_3CH_2CH_2CHCH_3 \qquad\qquad \textbf{E}$$
$$\underset{Br}{|}$$

(7) How may **E** be prepared?
Answer: **E** being an alkyl halide may be prepared from **F** via procedure III-1.

$$CH_3CH_2CH_2\overset{\overset{\textstyle H}{|}}{C}{=}CH_2 \qquad\qquad \textbf{F}$$

Note: **F** is the starting material we are requested to use.

(8) How may **D** be prepared?
Answer: **D** is an allylic alkyl bromide and thus can be prepared from **F** via the procedure III-3 as shown below:

$$\textbf{D} \xleftarrow{\text{NBS}} CH_3CH_2CH_2\overset{\overset{\textstyle H}{|}}{C}{=}CH_2 \qquad \textbf{F}$$

Thus, there are two procedures if we use **B** as our source of IV-c. Using the procedure which involves the fewest number of steps, the synthesis becomes

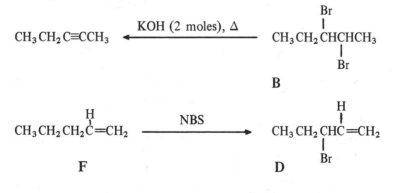

IV-d   Starting with

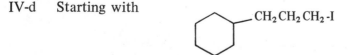

prepare

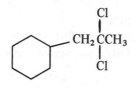

Let us ask some questions concerning IV-d.

(1) What kind of molecule is IV-d?
Answer: IV-d is a geminal alkyl dichloride (1-cyclohexyl-2,2-dichloropropane).

(2) Do we have to add carbon atoms?
Answer: No, because both the starting material and IV-d contain the same number of carbons.

(3) What procedure is available for the preparation of geminal dihalides?
Answer: An extension of procedure III-1 using an alkyne in place of an alkene will permit the formation of a geminal dihalide.

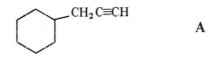

To form IV-d we need the alkyne **A**

(4) How may **A** be formed?
Answer: Since **A** is a terminal acetylene and we do not have to add carbon atoms, procedure IV-1 would be the easiest route. Thus, we need the dihalide **B** or **C**.

Let us use **B**, the reader may use **C** as an exercise.

(5) How may **B** be formed?

Answer: Since **B** is a vicinal dibromide, the easiest way of forming **B** is to add Br_2 across olefin **D** via procedure III-4.

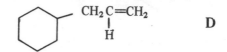

(6) How may **D** be prepared?

Answer: **D** is a terminal olefin and may be prepared by treatment of **E** with base via procedure II-1.

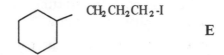

Note that **E** is the required starting material. Thus, the entire synthesis of IV-d is

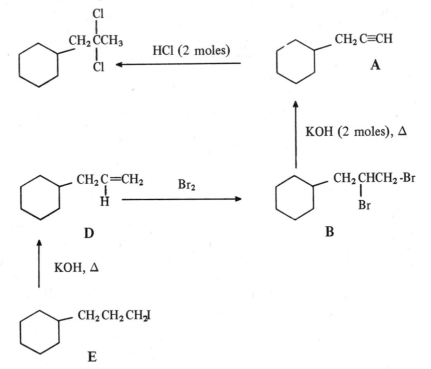

IV-e   Starting with

$$(CH_3)_3CCH_2CHCH_3$$
$$\underset{Br}{|}$$

prepare

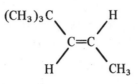

Let us ask some questions concerning IV-e.

(1) What kind of molecule is IV-c?
Answer: IV-e is a *trans*-olefin (*trans*-4,4-dimethyl-2-pentene).

(2) Do we have to add carbon atoms?
Answer: No, since the required starting material and IV-e each contain the same number of carbons.

(3) How can we prepare a *trans*-olefin?
Answer: Procedure II-3b which reduces an alkyne with Na and liquid NH_3

To prepare IV-e we need alkyne **A**.

$$(CH_3)_3CC{\equiv}CCH_3 \qquad\qquad \textbf{A}$$

(4) How may **A** be prepared?
Answer: Since carbon atoms do not have to be added, **A** can be prepared by treatment of the dihalides **B**, **C** or **D** with KOH via procedure IV-1.

Br	Br	Br
$(CH_3)_3CCCH_2CH_3$	$(CH_3)_3CCH_2CCH_3$	$(CH_3)_3CCHCHCH_3$
Br	Br	Br
B	**C**	**D**

We will use **D** leaving the preparations of **B** and **C** as an exercise for the reader.

(5) How may **D** be prepared?
Answer: **D** is a vicinal dibromide and may easily be formed

by the addition of Br_2 to an olefin. Thus, olefin **E** is needed.

$$(CH_3)_3 \underset{\underset{H}{|}}{C} C=CHCH_3 \qquad \textbf{E}$$

Although **E** has the correct structure of IV-e, it does not have the specified stereochemistry.

(6) How may **E** be prepared?
Answer: Since **E** is an olefin, it can readily be prepared by procedure II-1. Thus, either **F** or **G** would yield **E** upon treatment with base.

$$(CH_3)_3 C\underset{\underset{X}{|}}{C}HCH_2 CH_3 \qquad\qquad (CH_3)_3 CCH_2 \underset{\underset{X}{|}}{C}HCH_3$$

$$\textbf{F} \qquad\qquad\qquad\qquad \textbf{G}$$

Substituting Br for X we can readily see **G** is the required starting material. Thus the synthesis of IV-e becomes

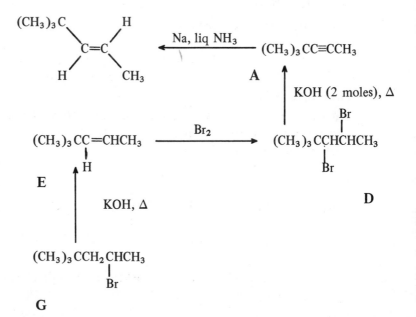

UNWORKED SYNTHESES

Using the same restrictions as stated for the syntheses of IV-a and IV-b prepare the following:

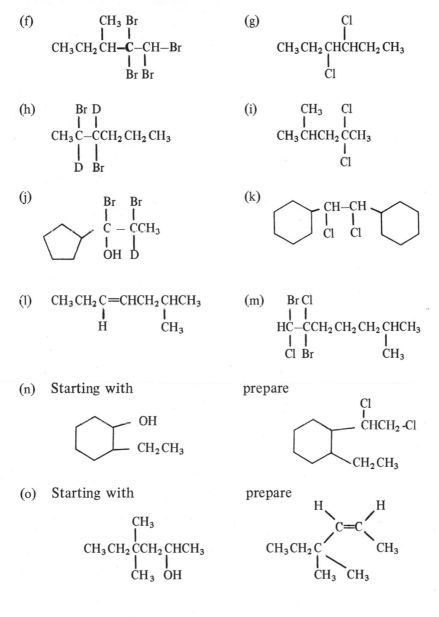

(f)

$$CH_3CH_2\underset{\underset{Br}{|}}{\overset{\overset{CH_3}{|}}{CH}}\!\!-\!\!\underset{\underset{Br}{|}}{\overset{\overset{Br}{|}}{C}}\!\!-\!\!CH\!\!-\!\!Br$$

(g)

$$CH_3CH_2\underset{\underset{Cl}{|}}{\overset{\overset{Cl}{|}}{CH}}CHCH_2CH_3$$

(h)

$$CH_3\underset{\underset{D}{|}}{\overset{\overset{Br}{|}}{C}}\!\!-\!\!\underset{\underset{Br}{|}}{\overset{\overset{D}{|}}{C}}CH_2CH_2CH_3$$

(i)

$$CH_3\overset{\overset{CH_3}{|}}{CH}CH_2\underset{\underset{Cl}{|}}{\overset{\overset{Cl}{|}}{C}}CH_3$$

(j)

(k)

(l) $$CH_3CH_2\underset{\underset{H}{\downarrow}}{C}\!\!=\!\!CHCH_2\overset{}{\underset{\underset{CH_3}{|}}{CH}}CH_3$$

(m) $$H\overset{}{\underset{\underset{Cl\ Br}{|\ |}}{C}}\!\!-\!\!CCH_2CH_2CH_2\overset{}{\underset{\underset{CH_3}{|}}{CH}}CH_3$$

(n)   Starting with   prepare

(o)   Starting with   prepare

CHAPTER V

AROMATIC ELECTROPHILIC SUBSTITUTION

In this chapter we shall discuss the syntheses of molecules containing aromatic rings. The compounds to be synthesized will contain the following substituents, halogens, nitro, sulfonic acid and alkyl groups. The synthesis of aromatic ketones employing the Friedel-Crafts acylation reaction will be discussed in Chapter X (Procedure X-6). The mechanistic implications will not be discussed in detail and the reader is referred to an organic text for this information before he begins these syntheses.

Tables I and II list the activation and deactivation as well as the directional properties of the various functional groups encountered in electrophillic aromatic substitution.

Table I lists various derivatives of benzene along with their electric moments which are a good measure of the ability of the various functional groups to donate or withdraw electron density from the ring. Negative signs (−) in front of the electric moment means that the group withdraws electron density from the benzene nucleus. A positive sign (+) means that the group donates electron density to the aromatic ring.

RULES FOR AROMATIC ELECTROPHILIC SUBSTITUTION

The following rules must be completely understood if one is going to be successful in performing electrophilic substitution reactions on aromatic rings already carrying substituents.

Table I. Electric Moments for Monosubstituted Benzenes

Group A C_6H_5-A	Electric Moment of C_6H_5-A	Direction of Moment C_6H_5-A or C_6H_5-A
$N(CH_3)_2$	1.68	
NH_2	1.53	
OH	1.45	
OCH_3	1.38	
$C(CH_3)_3$	0.83	+
$CH(CH_3)_2$	0.79	
CH_2CH_3	0.59	
CH_3	0.36	
H	0.0	
COOH	1.6	
F	1.6	
Cl	1.7	
Br	1.8	
CH_2Cl	1.8	
$COOC_2H_5$	1.9	
$CHCl_2$	2.0	
CCl_3	2.1	−
CHO	2.8	
$COCH_3$	3.0	
NO	3.1	
SO_3H	3.8	
NO_2	4.3	
CN	4.4	

Reprinted with permission from *Textbook of Organic Chemistry*, C. R. Noller, 3rd ed. (W. B. Saunders Co., 1966), p. 380.

(1) Groups already on the ring, which donate electron density to the aromatic ring, activate the ring toward electrophilic substitution relative to benzene.

(2) Groups already on the ring, which withdraw electron density from the ring, deactivate the ring toward further electrophilic substitution relative to benzene. (See Table I; $-N(CH_3)_2$ is the most activating group which $-CN$ is one of the most deactivating groups.)

(3) Groups which are considered to be activating groups, normally cause substitution to occur in the position's *ortho* (o) or *para* (p) to the activating group.

Table II. Relative Amounts of *ortho*, *para*, and *meta* Isomers Formed in the Nitration of Monosubstituted Benzenes

Group Present in Ring	Isomers Formed on Nitration (percentage)			
	ortho	*para*	*o + p*	*meta*
OH*	40	60	100	0
$CH(CH_3)_2$	14	86	100	0
CH_2CH_3	55	45	100	0
F	12	88	100	trace
Cl	29.6	69.5	99.1	0.9
Br	36.5	62.4	98.2	1.1
I	38.4	59.8	98.2	1.8
$NHCOCH_3$	19	79	98	2
CH_3	59	37	96	4
$C(CH_3)_3$	12	80	92	8
$CH_2COOC_2H_5$	42	47	89	11
$CH_2CH_2NO_2$	35	52	87	13
CH_2Cl	32	52	84	16
CH_2NO_2			67	33
$CHCl_2$	23	43	66	34
$[NH_3]^+$	1	52	53	47
$COCH_3$	45	0	45	55
CCl_3	7	29	36	64
$CONH_2$	27	3	30	70
$COOC_2H_5$	28	4	32	68
SO_3H	21	7	28	72
CHO	19	9	28	72
COOH	19	1	20	80
CN	17	2	19	81
NO_2	7	trace	7	93
SO_2CH_3	trace	trace	0	100
$[N(CH_3)_3]^+$	0	0	0	100

Reprinted with permission from *Textbook of Organic Chemistry*, C. R. Noller, 3rd ed. (W. B. Saunders Co., 1966), p. 382.

(4) Groups, which are considered to be deactivating groups cause substitution to occur in the position's *meta* (m) to the deactivating group (exceptions are the halogens which are strongly *o* and *p* directors).

Table II lists the relative amounts of *o*, *m* and *p* substitution products arising from the mono-nitration of monosubstituted benzenes.

(5) If an aromatic ring contains two activating groups *ortho* or *para* to one another, electrophilic substitution will take place *para* or *ortho* to the more activating group.

(6) If an aromatic ring contains an activating group and a de-activating group, the position of electrophilic substitution is controlled by the activating group.

PROCEDURES FOR AROMATIC ELECTROPHILIC SUBSTITUTION

A. Formation of alkyl substituted benzenes (Friedel-Crafts Reaction).

V-1* Use of an alkyl halide and a Lewis Acid ($AlCl_3$, $FeCl_3$, $SnCl_4$, BF_3).

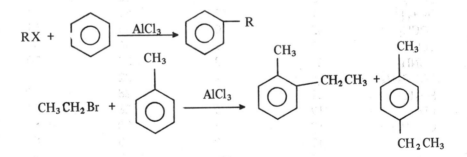

Note: This reaction will proceed only if the ring is not substituted with a *meta*-director group or an $-NR_2$ function.

---

*If there is only a *meta* directing group on the ring, the above reactions will fail. *e.g.* NO_2

$$CH_3CH_2CH_2\text{-Br, } AlCl_3 \longrightarrow \text{No Reaction}$$

V-2* Use of an olefin and Lewis Acid ($AlCl_3$, H_2SO_4).

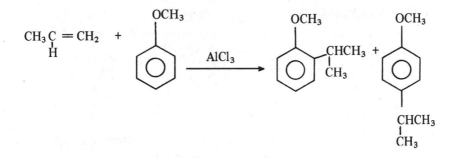

V-3* Use of an alcohol and H_2SO_4.

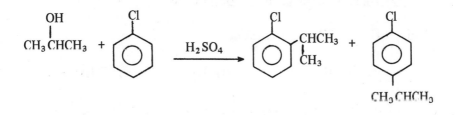

B. Formation of aryl halides.

V-4 Use of X_2 + FeX_3 where X = Cl and Br.
To insert a Cl or Br into a ring

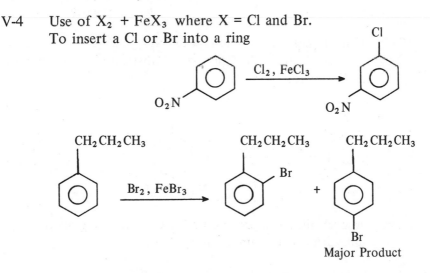

Major Product

V-5   Use of I_2 + HNO_3 or I_2 + HgO
      To insert an I into a ring

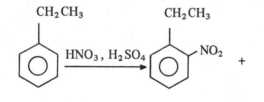

Major Product

HgO may be substituted for HNO_3 if there are acid sensitive groups in the molecule.

C.   Formation of aryl nitro compounds

V-6   Use of HNO_3 + H_2SO_4

CH$_2$CH$_3$ HNO$_3$, H$_2$SO$_4$ →   ortho NO$_2$ +   para NO$_2$

Major Product

V-7   Use of acetyl nitrate (CH_3COONO_2)
      Acetyl nitrate yields the *ortho* substituted nitro compound as the major product.

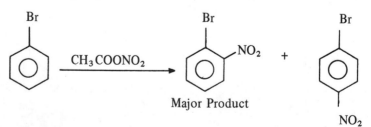

Major Product

Acetyl nitrate is prepared *in situ* by reacting acetic anhydride with fuming nitric acid.

D.   Formation of aromatic sulfonic acids

V-8   Use of H_2SO_4

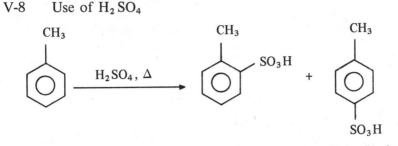

Major Product

V-9   Use of Chlorosulfonic Acid ($ClSO_3H$)

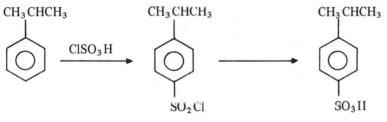

Major Product

E.   Chloromethylation of aromatic rings

V-10   Reaction of aromatic rings with $ZnCl_2$, CH_2O (formaldehyde) and HCl.

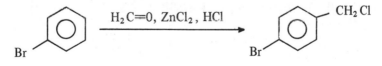

This reaction proceeds if the ring is substituted with
alkyl, halo or ether groups.

WORKED SYNTHESES

Starting with benzene (unless otherwise specified) prepare the
following compounds. You may use any monosubstituted acyclic
aliphatic organic molecules containing no more than four carbons
and any inorganic reagents.

V-a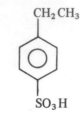
CH₂CH₃

SO₃H

Let us ask some questions concerning V-a.

(1) What kind of a molecule is V-a?
Answer: V-a is an aromatic sulfonic acid which contains an ethyl group *para* to the acid group (*p*-ethylbenzenesulfonic acid).

(2) Do we have to add carbon atoms?
Answer: Yes, because we are permitted to start with benzene and we have to add the ethyl group.

(3) What procedures are available for adding carbon atoms to an aromatic nucleus?
Answer: Procedures V-1, V-2 and V-3.

(4) What procedures are available for placing a sulfonic acid group on a ring?
Answer: Procedures V-8 and V-9.

(5) What is the order for placing the substituents on the aromatic ring?
Answer: The *para* arrangement of the groups on V-a gives us the main clue to begin the synthesis. The sulfonic acid group is a *meta* directing group (Table II). The ethyl group is an *ortho, para* directing group. Thus, the ethyl group must be on the molecule before the sulfonic acid group is placed on the molecule.
(Since we are working the problem backwards from the desired product, we must reverse the steps mentioned in the above paragraph, *i.e.*, the sulfonic acid group will be put on the ring before the ethyl group as shown below.)

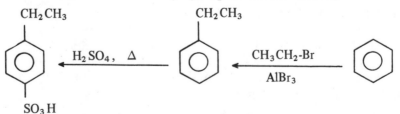

V-b

Let us ask some questions concerning V-b.

(1) What kind of molecule is V-b?
Answer: V-b is a chlorosubstituted nitrobenzene
(*o*-chloronitrobenzene).

(2) Do we have to add carbon atoms?
Answer: No, since all the carbon atoms in the molecule are
part of the benzene nucleus.

(3) What procedures are available for placing a $-NO_2$ group on
a ring?
Answer: Procedure V-6 and V-7, the latter yielding more
ortho substitution product.

(4) What procedure is available for placing a Cl on an aromatic
ring?
Answer: Procedure V-4.

(5) What is the order for placing the substituents on the aromatic
ring?
Answer: The *ortho* relationship of the Cl and $-NO_2$ pro-
vides the main clue to solving this problem. The Cl is an
ortho para directing group while the $-NO_2$ is a *meta* director.
This means that the Cl must be on the ring before the $-NO_2$.
(Since we are working the problem backwards from the desired starting material, we must reverse the order of group
placement on the ring as mentioned in the previous sentence
as shown below.)

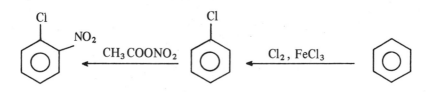

V-c

Let us ask some questions concerning V-c

(1) What kind of molecule is V-c?
   Answer: V-c is dihaloaromatic molecule (*p*-chloroiodobenzene).

(2) Do we have to add carbon atoms?
   Answer: No, since all the carbon atoms in the molecule are part of the benzene nucleus.

(3) What procedure is available for placing an I on an aromatic ring?
   Answer: Procedure V-5.

(4) What procedure is available for placing a Cl on an aromatic ring?
   Answer: Procedure V-4.

(5) What is the order for placing the substituents on the aromatic ring?
   Answer: Since both substituents are halogens, they both direct an incoming electrophile *ortho* or *para*. Thus, it makes no difference which halogens is put on the aromatic ring first. Let us place the Cl on the ring last (first if we work the problem backwards).

Thus, the synthesis of V-c becomes

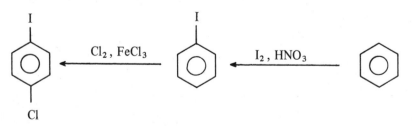

V-d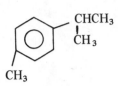

Let us ask some questions concerning V-d.

(1) What kind of a molecule is V-d?
Answer: V-d is a dialkyl aromatic molecule
(*p*-isopropyltoluene).

(2) Do we have to add carbon atoms?
Answer: Yes, the methyl group and the isopropyl group
require the addition of four carbons.

(3) What procedure is available for placing a $-CH_3$ group on an
aromatic ring?
Answer: Procedure V-1.

(4) What procedures are available for placing an $(H_3C)_2$ CH-
group on an aromatic ring?
Answer: Procedure V-1 which would require $CH_3 \overset{X}{CH}$-CH_3,
procedure V-2 which would require $CH_3CH=CH_2$, and
procedure V-3 which would require $CH_3 \underset{OH}{CH}$-CH_3.

(5) What is the order for placing the groups on the aromatic ring?
Answer: Since both substituents are *ortho para* directors,
we can put either group on the ring first. Let us put the —
$-CH_3$ group on the ring in the last step and use procedure
V-3 for placing the isopropyl group on the ring. Working
backwards stepwise the synthesis of V-d becomes

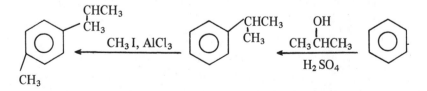

Reversing the above steps would also be a correct method
for the synthesis of V-d.

V-e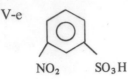

NO₂   SO₃H

Let us ask some questions concerning V-e.

(1) What kind of molecule is V-e?
Answer: V-e is a nitrosubstituted benzenesulfonic acid (*m*-nitrobenzenesulfonic acid).

(2) Do we have to add carbon atoms?
Answer: No, because all the carbon atoms of V-e are part of the benzene nucleus.

(3) What procedure is available for placing a $-NO_2$ group on an aromatic ring?
Answer: Procedure V-6 or V-7.

(4) What procedures are available for placing a $-SO_3H$ group on a benzene ring?
Answer: Procedure V-8 and V-9.

(5) What is the order for placing the substituents on the aromatic ring?
Answer: Since both the $-NO_2$ group and the $-SO_3H$ group are *meta* directors the directional factor will not be important in determining which group to put on the molecule last. However, since the $-NO_2$ group is a stronger aromatic ring deactivator for electrophilic substitution than the $-SO_3H$ group (Table I), it should probably be put on the ring after the $-SO_3H$ group.

Thus, working backwards in a stepwise fashion, the synthesis of V-e becomes

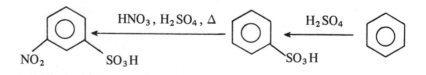

NO₂   SO₃H   ← HNO_3, H_2SO_4, Δ ←   SO₃H ← H_2SO_4 ←

V-f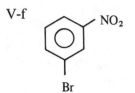

Let us ask some questions concerning V-f.

(1) What kind of molecule is V-f?
Answer: V-f is a nitro substituted aryl bromide (*m*-bromonitrobenzene).

(2) Do we have to add carbon atoms?
Answer: No, since all the carbons are part of the aromatic ring?

(3) What procedure is available for placing a $-NO_2$ group on an aromatic ring?
Answer: Procedure V-6.

(4) What procedure is available for placing a Br on an aromatic ring?
Answer: Procedure V-4.

(5) What is the order for placing the substituents on the aromatic ring?
Answer: The substitution pattern on the ring is important. The $-NO_2$ group is a *meta* director while the Br is an *ortho para* director with regard to an incoming electrophile. The *meta* relationship of the substituents on V-f dictates that the $-NO_2$ group should be placed on the ring before the Br.

Thus, working backwards in a stepwise manner, the synthesis of V-f becomes

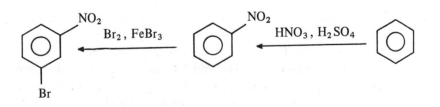

V-g

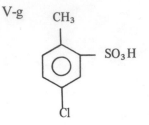

Let us ask some questions concerning V-g.

(1) What kind of a molecule is V-g?
Answer: V-g is a disubstituted benzenesulfonic acid (2-methyl-5-chlorobenzenesulfonic acid).

(2) Do we have to add carbon atoms?
Answer: Yes, the $-CH_3$ group requires the addition of one carbon.

(3) What procedure is available for placing a $-CH_3$ group on an aromatic ring?
Answer: Procedure V-1.

(4) What procedure is available for placing a Cl on an aromatic ring?
Answer: Procedure V-4.

(5) What procedures are available for placing a $-SO_3H$ group on an aromatic ring?
Answer: Procedures V-8 and V-9.

(6) What is the order for placing the substituents on the aromatic ring?
Answer: The $-CH_3$ and Cl are both *ortho para* directing groups. The $-SO_3H$ is a *meta* group. Since the $-CH_3$ group is *ortho* to the $-SO_3H$ group, the $-SO_3H$ should go on after the $-CH_3$. Because the $-SO_3H$ group is a strong acid, it may destroy some of the $FeCl_3$ which is the Lewis acid required to put the Cl on the ring. To obviate this happening, the Cl should be placed on the ring before the $-SO_3H$. Since the $-CH_3$ and Cl are as stated above *ortho para* directors, it does not matter a great deal which group is placed on the ring first. We shall place the $-CH_3$ group on the ring before the Cl.

Let us now synthesize V-g working backwards in a stepwise manner.

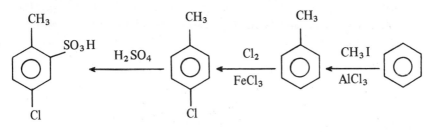

V-h

Let us ask some questions concerning V-h.

(1) What kind of molecule is V-h?
Answer: V-h is an alkyl and bromosubstituted nitrobenzene (2-ethyl-5-bromonitrobenzene).

(2) Do we have to add carbon atoms?
Answer: Yes, the two carbons of the ethyl group must be added.

(3) What procedures are available for placing an ethyl group on an aromatic ring?
Answer: Procedure V-1, V-2 and V-3.

(4) What procedure is available for placing a $-NO_2$ group on an aromatic ring?
Answer: Procedure V-6.

(5) What procedure is available for placing a Br on an aromatic ring?
Answer: Procedure V-4.

(6) What is the order for placing the substituents on the aromatic ring?
Answer: The $-C_2H_5$ group and Br are both *ortho para* directors while the $-NO_2$ is a strong *meta* director with regard to aromatic electrophilic substitution. The *ortho* relationship of the $-NO_2$ and $-C_2H_5$ groups dictates that

the $-C_2H_5$ be placed on the ring before the $-NO_2$ group. The *para* and *meta* relationship of the Br to the $-C_2H_5$ and $-NO_2$ groups respectively suggests that the Br may be placed on the ring before or after the $-NO_2$ group, and before or after the $-C_2H_5$ group. Thus the Br may be placed on the ring at any step. We shall place it on the ring after the $-C_2H_5$ group.

Using procedure V-3 as the source of the ethyl group the synthesis of V-h working backwards becomes

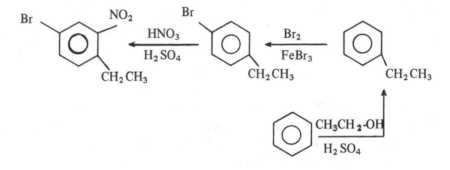

V-i

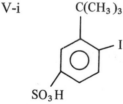

Let us ask some questions concerning V-i.

(1) What kind of molecule is V-i?
Answer: V-i is an alkyl and iodine substituted benzenesulfonic acid (3-*t*-butyl-4-iodobenzenesulfonic acid).

(2) Do we have to add carbon atoms?
Answer: Yes, the *t*-butyl group has to be added.

(3) What procedures are available for placing a *t*-butyl group on an aromatic ring?
Answer: Procedures V-1, V-2 and V-3.

(4) What procedure is available for placing an I on an aromatic ring?

Answer: Procedure V-5.

(5) What procedures are available for placing a $-SO_3H$ group on an aromatic ring
Answer: Procedure V-8 and V-9.

(6) What is the order for placing the substituents on the aromatic ring?
Answer: The I and the $-C(CH_3)_3$ group are both *ortho para* directors while the $-SO_3H$ group is a *meta* director. Since an *ortho para* director will control the substitution position of an incoming electrophile even if a *meta* director is also present on the ring, we shall concentrate on the I and $-C(CH_3)_3$ groups. The $-C(CH_3)_3$ group, because of its large size (bulk), presents us with a steric problem, which we have not previously encountered. Because of its large size the $-C(CH_3)_3$ normally goes to the *para* position of aromatic ring already substituted with another *ortho para* directing group. However, if the *para* position is blocked because of the presence of another group, it will then go to the *ortho* position. Thus, in order to get the $-C(CH_3)_3$ *ortho* to the I, it should be placed on the ring last. The *para* relationship of the I and $-SO_4H$ indicates that the I should be placed on the ring before the $-SO_3H$.

Thus, working backwards in a stepwise manner, the synthesis of V-i becomes

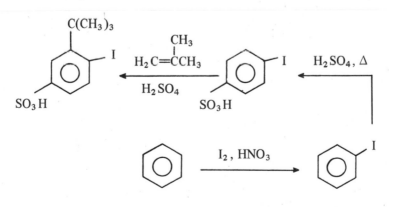

UNWORKED SYNTHESES

Synthesize the following using the restrictions put forth at the beginning of the worked syntheses section.

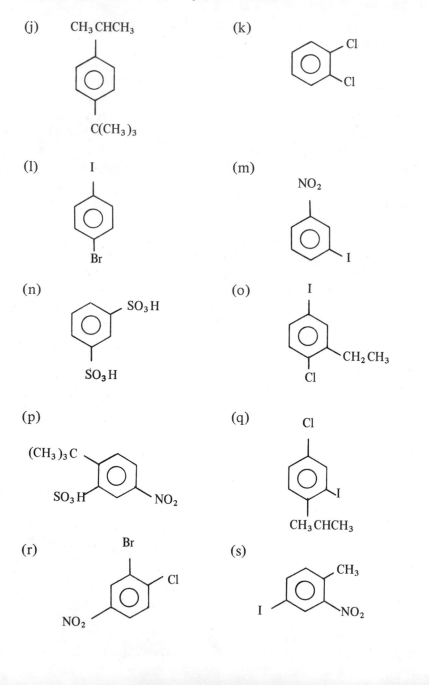

(j) CH_3CHCH_3

C(CH$_3$)$_3$

(k) Cl Cl

(l) I Br

(m) NO$_2$ I

(n) SO$_3$H SO$_3$H

(o) I CH$_2$CH$_3$ Cl

(p) (CH$_3$)$_3$C SO$_3$H NO$_2$

(q) Cl I CH$_3$CHCH$_3$

(r) Br Cl NO$_2$

(s) CH$_3$ I NO$_2$

CHAPTER VI

ALCOHOLS

There are many reactions in the chemical literature which generate an alcohol function. This book, in order to maintain brevity, will list only four of these procedures. Once again, it is emphasized that the student refer to lecture notes and organic texts for other alcohol preparations, which may be just as suitable or better for the synthesis of a specific alcohol.

PROCEDURES FOR SYNTHESIZING ALCOHOLS

VI-1*   Mercuric acetate [$Hg(OAc)_2$] and sodium borohydride ($NaBH_4$), Markovnikov addition of H_2O.

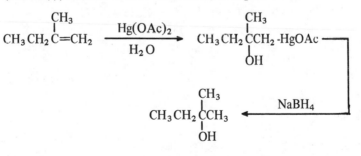

Note: This preparation requires two separate steps. The overall result of this procedure is that water which may be considered polarized in the following way

$$H \overset{\sigma^+}{—} OH^{\sigma^-}$$

---

*Neither VI-1 nor VI-2 react with aromatic rings.

adds across the olefin linkage to yield the Markovnikov addition product.

VI-2* Hydroboration (Diborane-B_2H_6), anti-Markovnikov addition of H_2O.

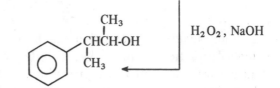

Once again this is a two-step reaction. If the overall reaction is considered the addition of a molecule of H_2O to the olefinic linkage, the alcohol which results is the anti-Markovnikov product.

VI-3 Reduction of carbonyl compounds

(a) Catalytically
Catalysts used (Ni, Pt, Pd, $CuO-Cr_2O_3$)
Aldehydes yield primary alcohols

$$CH_2CH_2\overset{\overset{O}{\parallel}}{C}H \xrightarrow{\ H_2,\ Pt\ } CH_3CH_2CH_2\text{-OH}$$

Ketones yield secondary alcohols

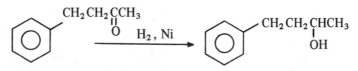

Note: If there are other groups on the molecule susceptible to hydrogenation, they may also be reduced if Ni, Pt or Pd are used. $CuO-Cr_2O_3$ (copper chromite) then would be the catalyst of choice since it is quite specific for the reduction of the carbonyl group.

(b) Chemically.

Reducing agents are lithium aluminum hydride (LiAlH₄) and NaBH₄.

As in the catalytic method above, other functional groups susceptible to reduction may interfere. In that case NaBH₄ is the only choice since it is highly specific for the reduction of aldehydes and ketones.

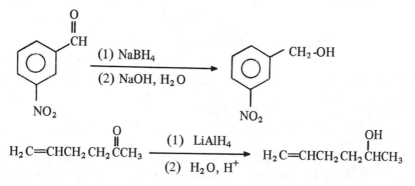

VI-4 Grignard Reagent

The most useful procedure for the preparation of alcohols is the use of organometallic compounds known as Grignard reagents. The most frequently used are those Grignard reagents which contain Mg as the metal. The Grignard reagent is commonly written RMgX, where R is an organic moiety and X is a halogen (Cl, Br or I).

The utility of the Grignard reagent as a synthetic tool for the production of alcohols can best be demonstrated if one classifies the type alcohol to be synthesized as primary, secondary or tertiary.

(a) Preparation of primary alcohols (R-CH₂OH)

$$RMgX + H_2C{=}0 \longrightarrow RCH_2-OMgX \xrightarrow{\ H_2O,\ H^+\ } RCH_2-OH$$

$$RMgX + \overset{O}{\overset{\diagup \diagdown}{CH_2-CH_2}} \longrightarrow RCH_2CH_2-OMgX \xrightarrow{\ H_2O,\ H^+\ } RCH_2CH_2-OH$$

Formaldehyde (CH₂O) may be used in the preparation of any primary alcohol, while ethylene oxide $(\overset{O}{\overset{\diagup \diagdown}{CH_2\,CH_2}})$ may

be used only if the desired alcohol has the structure
$-CH_2 CH_2 OH$.

(b) Preparation of secondary alcohols $(R-\overset{R'}{\underset{H}{C}}OH)$

A secondary alcohol requires a Grignard reagent and an aldehyde.

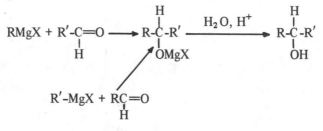

Note either R or R′ may be attached to the required aldehyde function, and likewise R or R′ may contain the Grignard reagent. Thus, there are two ways of arriving at the same alcohol.

(c) Preparation of tertiary alcohols

A tertiary alcohol requires a Grignard reagent and a ketone $(R-\overset{O}{\underset{}{C}}-R)$. There are three possible ways of arriving at the same alcohol.

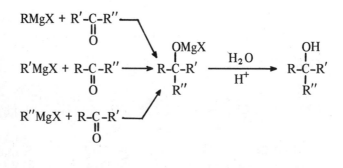

Note: If the following groups are present on either the Grignard reagent or the carbonyl containing compound, procedure VI-4 often produces an unsatisfactory yield of of the desired alcohol and other procedures must be used. The interfering groups are $-OH$, $-NH_2$, $-CO_2H$, $-CO_2R$, $-C\equiv N$, $-NO_2$.

WORKED SYNTHESES

Starting with the compound on the left, prepare the compound on the right. You may use any inorganic reagents you wish.

VI-a

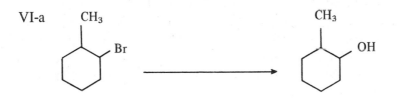

Let us ask some questions about the desired alcohol VI-a.

(1) What kind of alcohol is VI-a?
Answer: VI-a is a cyclic alcohol (2-methylcyclohexanol).

(2) Do we have to add carbon atoms?
Answer: No, because the prescribed starting material and VI-a each contain seven carbons.

(3) What procedures are available for placing an $-OH$ on a molecule without adding unwanted carbon atoms?
Answer: Procedure VI-1, VI-2 and VI-3. Since procedure VI-3 requires the use of a carbonyl group, which we have not yet discussed the preparation of, let us use procedure VI-1 or VI-2.

(4) What starting materials are required which via procedures VI-1 and VI-2, VI-a will be the major product?
Answer: Procedure VI-1 would require olefin **A**, while procedure VI-2 would require olefin **B**.

It should be noted that using procedure VI-1, VI-a will form in only about 50% yield, while using olefin **B** and procedure VI-2, VI-a will be the sole product. Thus, we shall use olefin **B**.

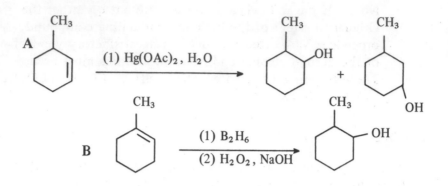

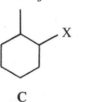 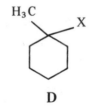

(5) What procedures are available for the preparation of **B**?
Answer: Since **B** is an olefin, we must refer to Chapter II.
Keeping in mind that the prescribed starting material is an
alkyl halide, procedure II-1 would be the method of choice.
This would require an alkyl halide with the structure of **C**
or **D**.

If X = Br, we can easily see that **C** is the structure of the
required starting material.

Thus, the synthesis of VI-a becomes

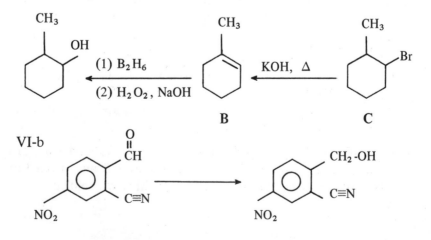

Let us ask some questions concerning the alcohol VI-b.

(1) What kind of molecule is VI-b?
Answer: VI-b is a substituted benzyl alcohol (2-cyano-4-nitrobenzyl alcohol).

(2) Do we have to add carbon atoms?
Answer: No, because VI-b and the prescribed starting material each have the same number of carbons.

(3) What procedures are available for the synthesis of alcohols but do not add carbon atoms?
Answer: Procedures VI-1, VI-2 and VI-3. Since we have an aldehyde function ($-\overset{\overset{\text{O}}{\|}}{\text{C}}$-H) present in the required starting material, let us use procedure VI-3.

(4) Are the other groups on the ring susceptible to reduction?
Answer: Yes, the $-NO_2$ group can be reduced to the $-NH_2$ group while the $-C\equiv N$ can be reduced to the $-CN_2NH_2$ group.

(5) What reagent will reduce the -CHO function and leave the $-NO_2$ and $-C\equiv N$ groups intact?
Answer: $NaBH_4$.

The synthesis of VI-b now reduces to a simple one-step reaction once the correct reducing agent is found.

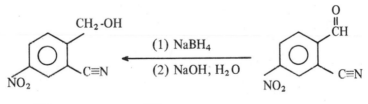

VI-c
$$\underset{\underset{\text{Cl}}{|}}{CH_3\,CHCHCH_3} \xrightarrow{\quad} \underset{\underset{\text{OH}}{|}}{\overset{\overset{\text{CH}_3}{|}}{CH_3\,CCH_2\,CH_3}}$$

Let us ask some questions concerning the alcohol VI-c.

(1) What kind of molecule is VI-c?
Answer: VI-c is a tertiary aliphatic alcohol (2-methyl-2-butanol).

(2) Do we have to add carbon atoms?

Answer: No, because the designated starting material and VI-c contain the same number of carbons.

(3) What procedures are available for placing a –OH on a molecule without adding carbon atoms?
Answer: Procedures VI-1, VI-2 and VI-3. Since we are not permitted to start with a carbonyl containing compound and we do not know how to incorporate a carbonyl group into a molecule, let us concentrate on procedures VI-1 and VI-2 which require olefin precursors.

(4) What are the structures of the olefins which through the use of procedures VI-1 and VI-2 will yield VI-c as the major product?
Answer: Procedure VI-1 requires olefin **A** or **B**

A   B

while procedure VI-2 cannot be used to form VI-c because VI-c is a tertiary alcohol.

(5) How can **A** be formed?
Answer: Since **A** is an olefin, procedures II-1 and II-2 would suffice. Since we are requested to start with an alkyl chloride, let us use procedure II-1. Thus an alkyl chloride of structure **C** or **D** would yield **A**.

C   D

Note: **D** is the structure of the required starting material. Thus, the synthesis of VI-c becomes

A   D

For the rest of the alcohol syntheses you may use any monosubstituted cyclic molecule, any monosubstituted acyclic organic molecule containing no more than four carbons and any inorganic reagents.

VI-d

Let us ask some questions concerning VI-d.

(1) What kind of molecule is VI-d?
Answer: VI-d is a secondary aromatic alcohol [1-(*p*-chloro-phenyl)butanol].

(2) Do we have to add carbon atoms?
Answer: Since we are limited to starting with monosubstituted cyclic molecules we have the option of adding the carbons to the chlorinated ring or adding the Cl to the aromatic alcohol. In the lab it is easier to do the former, so we shall proceed along that line. Thus, we have to add carbons.

(3) What procedure is available for adding carbons to a molecule and yields an alcohol as the product?
Answer: Procedure VI-4, which utilizes a Grignard reagent.

(4) How can VI-d be broken up (which C—C bond cleaved) so that the structures of the required carbonyl compound and the Grignard reagent can be deduced?
Answer: VI-d is a secondary alcohol. Thus, we need a Grignard reagent and an aldehyde. Being a secondary alcohol VI-d may be broken up in two ways, namely by cleaving either of the bonds connecting the carbinol carbon to the adjacent carbons as indicated below.

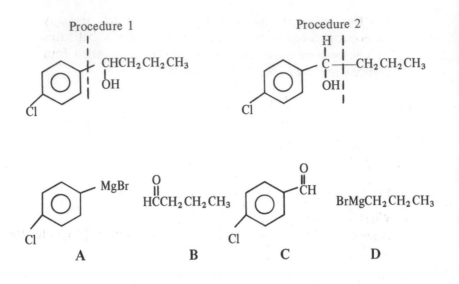

Procedure 1

Procedure 2

A   B   C   D

Since we can start with the aldehyde **B** which is needed from the procedure 1 bond cleavage, while the aldehyde from the procedure 2 bond cleavage requires preparations which we have not yet covered, let us cleave the C—C bond as indicated in procedure 1 above.

(5) How can **A** be prepared?

Answer: **A** can be prepared by treating

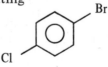

with Mg. It should be remembered that aromatic Br and I form Grignards reagents rather easily, while aromatic Cl form Grignards reagents with difficulty.

(6) How can **C** be prepared?

Answer: Procedure V-4, which is the bromination of chlorobenzene, a permitted starting material.

Thus, working backwards, the synthesis of VI-d becomes

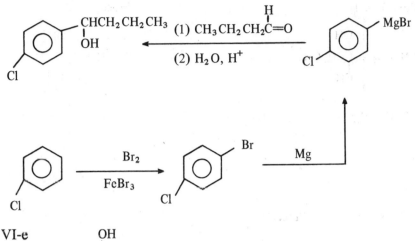

VI-e

$$CH_3CH_2\underset{\underset{CH_3}{|}}{\overset{\overset{OH}{|}}{C}}CH_2CH_3$$

Let us ask some questions concerning VI-e.

(1) What kind of molecule is VI-e?
Answer: VI-e is a tertiary aliphatic alcohol (3-methyl-3-pentanol).

(2) Do we have to add carbon atoms?
Answer: Since VI-e has six carbons and we are permitted to use molecules containing no more than four carbons, yes, we must add carbon atoms.

(3) What procedure is available for adding carbon atoms and yields an alcohol at the product?
Answer: Procedure VI-4, which utilizes a Grignard reagent.

(4) How can VI-e be broken up (which C—C bond cleaved) so that the structures of required carbonyl compound and the Grignard reagent can be deduced?
Answer: Since VI-e is a tertiary alcohol, there are three ways (three C–C–OH bonds) in which the molecule may be broken up into its component parts. The three ways are illustrated by procedures 1, 2 and 3 below.

Procedure 1

$$CH_3CH_2 \overset{|}{\underset{\underset{CH_3}{|}}{\overset{OH}{\underset{|}{C}}}}CH_2CH_3$$

CH_3CH_2MgBr $O{=}\overset{|}{\underset{\underset{CH_3}{|}}{C}}CH_2CH_3$

Procedure 2

$$CH_3CH_2 \overset{OH}{\underset{\underset{CH_3}{|}}{\overset{|}{C}}}CH_2CH_3$$

$CH_3CH_2 \overset{}{\underset{\underset{O}{\|}}{C}}CH_2CH_3$

CH_3MgI

Procedure 3

$$CH_3CH_2 \overset{OH}{\underset{\underset{CH_3}{|}}{\overset{|}{C}}}{-}CH_2CH_3$$

$CH_3CH_2\overset{}{\underset{\underset{CH_3}{|}}{C}}{=}O$ $BrMgCH_2CH_3$

Note that all three procedures require a Grignard reagent and a ketone. Note also that procedures 1 and 3 require the same Grignard reagent and ketone. Thus, our choice comes down to whether to use procedure 1 or procedure 2. If we use procedure 1, we can start with the needed ketone since it contains only four carbons. If we use procedure 2, we cannot start with the required ketone because it contains five carbons. Let us, therefore, use procedure 1 because it will require fewer steps.

The synthesis of VI-3 working backwards now becomes

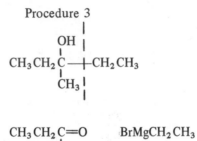

VI-f

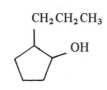

Let us ask some questions concerning VI-f.

(1) What kind of molecule is VI-f?
Answer: VI-f is a cyclic secondary alcohol (2-propylcyclo-pentanol).

(2) Do we have to add carbon atoms?
Answer: Since VI-f is a disubstituted cyclic molecule and it is easier to add carbons to an alcohol than to add an –OH to an alkane, yes, we will have to add carbons.

(3) Since we have to add carbons and the desired product is an alcohol, is it possible to use a Grignard reagent to produce VI-f?
Answer: Although in theory a Grignard reagent could be utilized to synthesize VI-f in one step, we do not currently have at hand the method of preparing the required halo aldehyde. A Grignard reaction can be used to prepare cyclic alcohols provided the Grignard reagent carbons are added to the carbinol carbon (C–OH). In VI-f the propyl group is attached to the carbon adjacent to the carbinol carbon.

(4) What other procedures are available for preparing alcohols?
Answer: Procedures VI-1, VI-2 and VI-3. Procedure VI-1 would require olefin A, procedure VI-2 would require olefin B and procedure VI-3 would require ketone C.

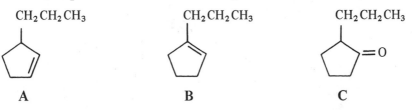

Olefin A would yield at best a 50% yield of the desired alcohol via procedure VI-1. Ketone C requires the preparation of a ketone which we have not yet covered. Olefin B would give exclusively VI-f via procedure VI-2 and thus this is the method of choice. Since **B** is disubstituted it must be synthesized.

(5) How can **B** be prepared?
Answer: Procedures II-1 and II-2. Keeping in mind that we still have to add the propyl group to the ring, which can best be accomplished through a Grignard reaction, let us use procedure II-2. This would require alcohols **D** or **E**.

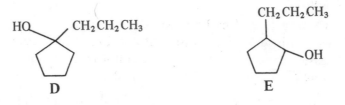

D   E

Since **E** is the molecule we are attempting to prepare, it would be ridiculous to use it. Thus **D** is the alcohol of choice.

(6) How can **D** be prepared?
Answer: **D** is a tertiary alcohol which can be prepared by using Grignard reagent and a ketone. Note that the alkyl group is attached to the carbinol carbon, thus this compound can readily be made by a Grignard reaction.

(7) How can **D** be broken up into the required ketone and Grignard reagent?
Answer: The best C–C bond to break in a cyclic aliphatic alcohol is the bond which attaches the alkyl substituent to the carbinol carbon as indicated below

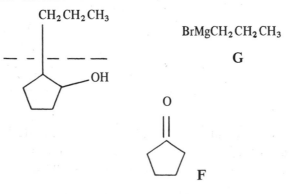

F is a permitted starting material, while **G** can be prepared by treating 1-bromopropane with Mg.

Working backwards in a stepwise manner, the synthesis of VI-f becomes

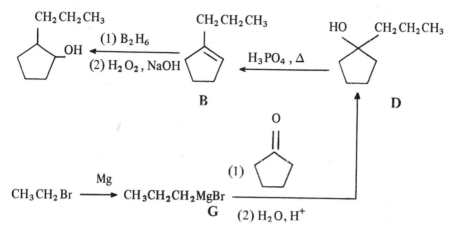

B

D

$$CH_3CH_2Br \xrightarrow{Mg} CH_3CH_2CH_2MgBr$$

G

(2) H_2O, H^+

VI-g

$$CH_3\overset{\overset{\displaystyle CH_3}{|}}{C}-\overset{\overset{\displaystyle}{|}}{C}HCH_2CH_2CH_3$$
$$\overset{|}{O}H \ \overset{|}{C}l$$

Let us ask some questions concerning VI-g.

(1) What kind of molecule is VI-g?
Answer: VI-g is an aliphatic vicinal tertiary chloroalcohol (2-methyl-3-chloro-3-hexanol).

(2) Do we have to add carbon atoms?
Answer: Yes, because VI-g contains seven carbons and we are permitted to start with acyclic molecules containing no more than four carbons.

(3) What procedure is available for the preparation of vicinal chloralcohols?
Answer: Procedure III-5 which would require olefin **A**

$$CH_3\overset{\overset{\displaystyle CH_3}{|}}{C}=CHCH_2CH_2CH_3$$

A

(4) How can **A** be prepared?
Answer: **A** can be prepared via procedures II-1 and II-2.
Keeping in mind that since we must add carbon atoms
during some stage of the synthesis and that the Grignard
reagent reaction will allow one to add carbons and yield
an alcohol as the product, let us use procedure II-2.

(5) What is the structure of the alcohol which would yield **A**
upon dehydration?
Answer: Alcohols **B** and **C** will yield **A** upon dehydration.

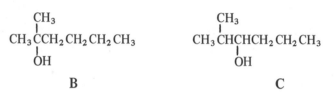

B   C

(6) How can **B** and **C** be prepared?
Answer: **B** is a tertiary alcohol and thus can be prepared by
the reaction of a Grignard reagent and a ketone. **B** can be
broken up in three ways, one of which is shown below to
reveal the components needed for the preparation of **B**.

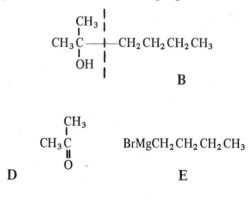

C is a secondary alcohol and can be synthesized by reaction
of a Grignard reagent and an aldehyde. **C**, thus, can be
broken up in two ways to reveal the components necessary
for the synthesis of **C**.

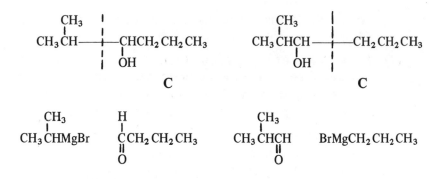

Note that all the Grignard reagents, aldehydes and ketone shown above do not contain more than four carbons. Thus, we may use **B** or **C** and any of their preparations illustrated above.

Using **B** as the precursor of **A**, let us now synthesize VI-g by working backwards from the desired starting material in a stepwise fashion.

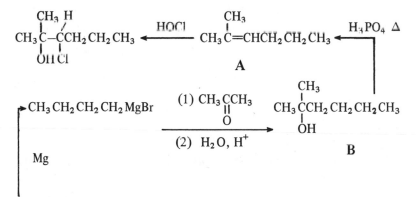

VI-h

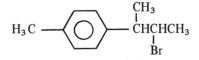

Let us ask some questions concerning VI-h.

(1) What kind of molecule is VI-h?

Answer: VI-h is a *para* disubstituted aryl aliphatic halide (2-bromo-3-*p*-tolylbutane).

(2) Do we have to add carbon atoms?
Answer: Yes, since we are permitted to start with monosubstituted ring compounds and monosubstituted aliphatic molecules containing no more than four carbons.

(3) What procedures are available for adding carbon atoms?
Answer: Procedures I-3 and VI-4. Since the use of procedure I-3 would require a vicinal dihalide, the yield of the desired product would be quite low. Thus we shall use the Grignard reagent procedure to add the necessary carbon atoms.

(4) What procedures are available for preparing an aliphatic bromide?
Answer: Procedures III-1 and III-2. Procedure III-1 would necessitate the synthesis of olefin **A**, while procedure III-2 would require olefin **B**.

$$H_3C - \text{\large\bigcirc} - \overset{\overset{\displaystyle CH_3}{|}}{C}HCH{=}CH_2 \qquad H_3C - \text{\large\bigcirc} - \overset{\overset{\displaystyle CH_3}{|}}{C}{=}CHCH_3$$

$$\textbf{A} \qquad\qquad\qquad\qquad \textbf{B}$$

Since it is normally easier to generate a C=C in conjugation with an aromatic ring, we shall use olefin **B** and procedure III-2.

(5) How can **B** be prepared?
Answer: **B** being an olefin can be prepared by procedures II-1 and II-2. Since we have already decided that carbon atoms are going to be added via a Grignard reaction which will yield an alcohol as the product, let us use procedure II-2, the dehydration of an alcohol, to prepare **B**. The structure of the alcohols which will yield **B** as the principle product are labelled **C** and **D** below.

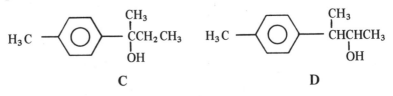

$$\textbf{C} \qquad\qquad\qquad\qquad \textbf{D}$$

Since we are going to have to connect the large aliphatic group to the ring at some stage of the synthesis, it will be easier if we select **C** in which the carbinol carbon is attached directly to the ring.

(6) How can **C** be prepared?
Answer: **C** being a tertiary alcohol can be prepared by the reaction of a Grignard reagent and a ketone via procedure VI-4.

(7) How can **C** be broken up so that the structures of the required Grignard reagent and ketone may be deduced?
Answer: Since **C** is a tertiary alcohol, there are three carbon-carbinol carbon bonds which may be cleaved to reveal the needed starting materials.

Cleavage 1

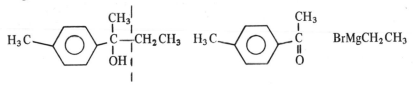

Cleavage 2

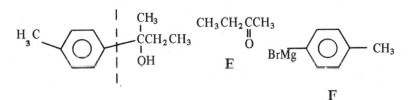

Cleavage 3

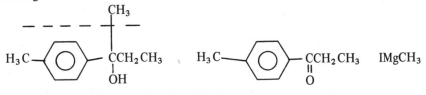

It can be readily seen that only in the first cleavage we do not have to synthesize the ketone. In the latter two cleavages, the synthesis of the ketone is required. Since we have not covered the synthesis of ketones, let us use the starting materials required by the first cleavage. (It should be noted that **C** may be produced equally as well if one starts with the ketones and Grignard reagents stipulated in the latter two cleavages).

(8) How can **F** be prepared?
Answer: **F** can be prepared from the corresponding *p*-bromotoluene **G**.

(9) How can **G** be prepared?
Answer: **G** can be prepared by the bromination of toluene via procedure V-4.

Thus working backwards in a stepwise fashion, the synthesis VI-h becomes

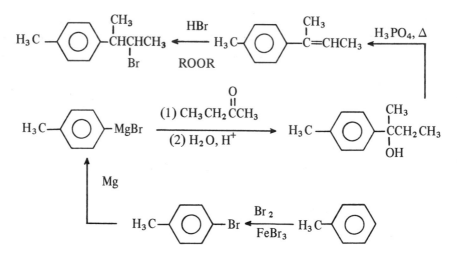

SYNTHESES PROBLEMS

Starting with the compound on the left, prepare the compound on the right. You may use any inorganic reagents you wish.

(j)

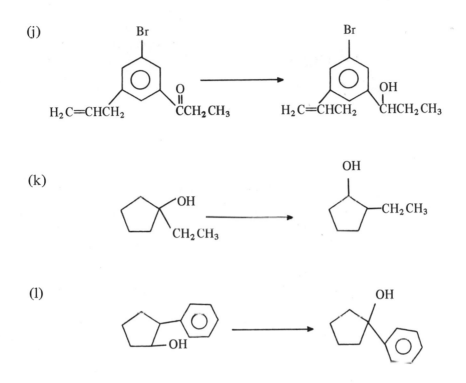

(k)

(l)

UNWORKED SYNTHESES

For the rest of the synthesis problems, you may use any mono-substituted cyclic organic molecule, any monosubstituted acyclic organic molecule containing no more than four carbons and any inorganic reagents.

(m) $(CH_3)_3CCH_2CH_2$-OH

(n) $CH_3CH_2\underset{\underset{\textstyle CH_2CH_3}{|}}{CH}$–OH

(o)

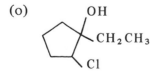

(p) CH_2OH

(q)

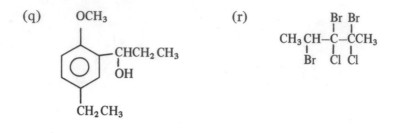

OCH₃

−CHCH₂CH₃
  |
  OH

CH₂CH₃

(r)

$$CH_3\,CH-C-CCH_3$$

Br Br
| |
C
| |
Br Cl Cl

(s)

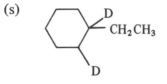

D
− CH₂CH₃

D

(t)

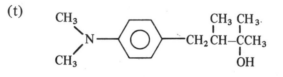

CH₃
   \
    N−
   /
CH₃

CH₃ CH₃·
 | |
−CH₂CH−CCH₃
         |
         OH

(u)

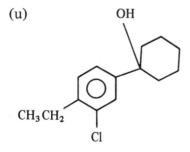

OH

CH₃CH₂

Cl

(v)

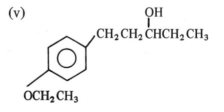

OH
 |
CH₂CH₂CHCH₂CH₃

OCH₂CH₃

CHAPTER VII

ETHERS

The ether linkage ($\geq$C–0–C$\leq$) is normally prepared by one of the three following procedures. The Williamson synthesis is the most general method, while the diazomethane procedure and the epoxide formation procedure are useful for specific types of ethers.

PROCEDURES FOR SYNTHESIZING ETHERS

VII-1 Williamson Synthesis—preparation of symmetrical and unsymmetrical ethers.
This reaction proceeds via an $S_N 2$ mechanism. Therefore, it is susceptible to the same complication that accompanies any $S_N 2$ reaction, namely elimination. To reduce the amount of elimination product, the alkyl halide should be either primary or secondary.

$$RO^- Na^+ + R'X \longrightarrow ROR'$$

$$CH_3 \underset{\underset{CH_3}{|}}{CH}O^- Na^+ + CH_3 CH_2 Br \longrightarrow CH_3 \underset{\underset{CH_3}{|}}{CH}OCH_2 CH_3$$

VII-2 Diazomethane—preparation of methyl ethers

$$ROH \xrightarrow{CH_2 N_2} ROCH_3$$

Alcohols normally require the use of a catalyst; HBF_4 is frequently employed for that purpose.

$$CH_3CH_2CH_2\underset{\underset{CH_3}{|}}{\overset{\overset{OH}{|}}{C}}CH_3 \xrightarrow[HBF_4]{CH_2N_2} CH_3CH_2CH_2\underset{\underset{CH_3}{|}}{\overset{\overset{OCH_3}{|}}{C}}CH_3$$

For phenols, no catalyst is required

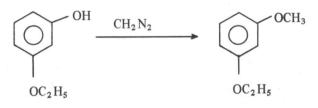

VII-3   Exposide formation

The most common method of preparing epoxides

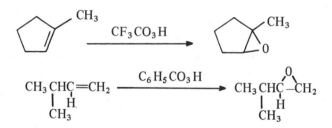

involves the use of peracids $(R-CO_3H)$. Trifluoroperacetic acid (F_3C-CO_3H) and peroxybenzoic acid $(C_6H_5CO_3H)$ are the most frequently used peracids.

WORKED SYNTHESES

Using only monosubstituted cyclic organic molecules, monosubstituted acyclic organic molecules containing no more than four carbons and any inorganic reagents prepare the following.

VII-a

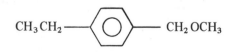

Let us ask some questions concerning VII-a.

(1) What kind of molecule is VII-a?
Answer: VII-a is a *para* disubstituted aromatic molecule containing an aliphatic methyl ether linkage (4-ethyl-α-methoxytoluene).

(2) Do we have to add carbon atoms?
Answer: Yes, since we are permitted to start with mono-substituted cyclic molecules and VII-a is a disubstituted molecule in which each substituent contains carbons.

(3) What procedures are available for preparing methyl ethers?
Answer: Procedures VII-1 and VII-2. Since procedure VII-2 is specific for methyl ethers, let us use that procedure.

(4) What is the structure of the alcohol which upon reaction with diazomethane will yield VII-a as the product?
Answer: The structure of the desired alcohol would be the molecule labelled A.

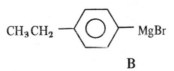

A

(5) How can **A** be prepared?
Answer: Since **A** is a primary alcohol, it can be prepared by reaction of the Grignard agent **B** and formaldehyde via procedure VI-4a.

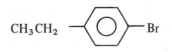

B

(6) How can **B** be prepared?
Answer: **B** can be prepared by reaction of the aryl halide **C** with Mg.

CH_3CH_2 ─⟨◯⟩─ Br

C

(7) How can C be prepared?

Answer: C can be prepared by reaction of ethylbenzene with Br_2 via procedure V-4. Ethylbenzene is a permitted starting material.

Thus the synthesis of VII-a, working backwards in a step-wise manner, becomes:

$$CH_3CH_2-\langle\bigcirc\rangle-CH_2OCH_3 \xleftarrow[HBF_4]{CH_2N_2} CH_3CH_2-\langle\bigcirc\rangle-CH_2OH \xleftarrow[(2)\ H_2O,\ H^+]{(1)\ H_2C=O}$$

A

$$CH_3CH_2-\langle\bigcirc\rangle \xrightarrow[FeBr_3]{Br_2} CH_3CH_2-\langle\bigcirc\rangle-Br \xrightarrow{Mg} CH_3CH_2-\langle\bigcirc\rangle-MgBr$$

C   B

VII-b

$$CH_3CH_2\overset{O}{\overset{/\backslash}{C}}-\underset{\underset{CH_3}{|}}{C}HCH_3$$

Let us ask some questions concerning VII-b.

(1) What kind of molecule is VII-b?

Answer: VII-b is an aliphatic epoxide (3-methyl-2,3-epoxy-pentane).

(2) Do we have to add carbon atoms?

Answer: Since VII-b contains six carbons and we are permitted to utilize monosubstituted organic molecules containing no more than four carbons, yet, we do have to add carbon atoms.

(3) What procedure is available for the preparation of epoxides?

Answer: Procedure VII-3 which would require an olefin and peroxybenzoic acid. The structure of the required olefin can be determined by placing the double bond between the carbons which contain the ether linkages in VII-b.

Thus olefin **A** is required.

$$CH_3 CH_2 \underset{\underset{CH_3}{|}}{C}=CHCH_3$$

A

(4) How can **A** be prepared?

Answer: **A** can be prepared via procedure II-1 or II-2. Keeping in mind that we have to add carbon atoms and that the Grignard reaction is an excellent way to accomplish this, let us use procedure II-2, the dehydration of an alcohol, which would be the product of the aforementioned Grignard reaction.

(5) What is the structure of the alcohol which upon dehydration would yield **A**?

Answer: Alcohols **B** and **C** upon treatment with $H_3 PO_4$ would yield **A**.

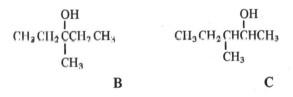

B   **C**

We shall select **B** as our choice although **C** can also be used. The use of **C** will be left to the student as an exercise.

(6) How can **B** be prepared?

Answer: **B** is a tertiary alcohol and thus can be prepared from a Grignard reagent and a ketone. **B** contains three carbon-carbinol carbon bonds which may be cleaved to permit the deduction of the structure of the required Grignard reagent and ketone.

$$CH_3 CH_2 \underset{\underset{CH_3}{|}}{\overset{\overset{OH}{|}}{C}}\text{---}CH_2 CH_3$$

We shall cleave the bond shown. This shows us that we need the ketone **D** and the Grignard reagent **E**. **D** is a permitted starting material and **E** can be prepared by reacting ethyl bromide and Mg.

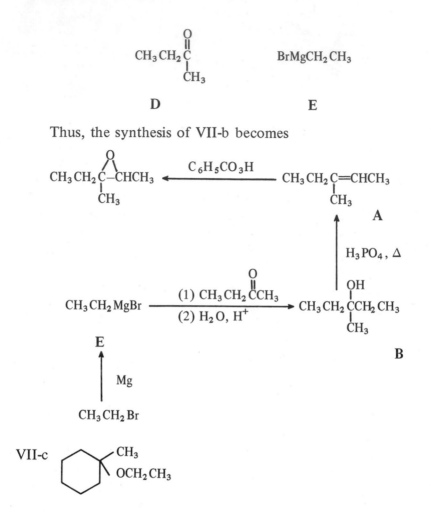

Thus, the synthesis of VII-b becomes

Let us ask some questions concerning VII-c.

(1) What kind of molecule is VII-c?
    Answer: VII-c is an unsymmetrical aliphatic ether (1-methyl-cyclohexyl ethyl ether).

(2) Do we have to add carbon atoms?
    Answer: Yes, because VII contains a ring containing two substituents each of which contains carbon atoms.

(3) What procedure is available for the preparation of an unsymmetrical ether?

Answer: Since neither of the alkyl groups is methyl in which case procedure VII-2 could be used, only procedure VII-1 is available.

(4) How can VII-c be broken up (which bonds cleaved) so that the starting materials for the synthesis of VII-c via procedure VII-1 may be deduced?

Answer: VII-c may be broken up by cleaving the carbon-oxygen ether linkage. Since there are two C–O ether bonds, there are two ways of breaking the molecule up as illustrated below.

Cleavage 1   Cleavage 2

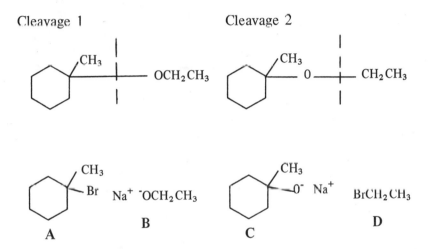

Cleavage 1 would require the tertiary Br, **A**, while cleavage 2 would require the primary Br, **D**. Since the Williamson synthesis works best if the alkyl halide is primary rather than tertiary, cleavage 2 is the method of choice.

(5) How can **C** be prepared?

Answer: **C** can be prepared by the reaction of Na with alcohol **E**

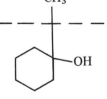

(6) How can E be prepared?
Answer: Since E is a tertiary alcohol, it can be prepared
from a Grignard reagent and a ketone. Being a tertiary
alcohol there are three possible ways of breaking up the
molecule by cleaving carbon-carbinol carbon bonds. As
stated before, with a cyclic alcohol the easiest C–C $\overset{OH}{\text{C}}$ bond
to cleave is the one between the alkyl substituent (CH$_3$)
and the ring as illustrated above. This would require

 and IMgCH$_3$ as the starting materials.

Cyclohexanone and CH$_3$I are permitted starting materials.
The synthesis of VII-c now becomes?

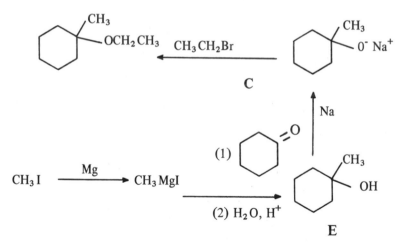

UNWORKED SYNTHESES

Prepare the following using the same instructions that were
stated before the worked syntheses section.

(d)   CH$_3$ CH$_3$ (e)
       | |
CH$_3$CH–O–CH$_2$CHCH$_3$

(f)

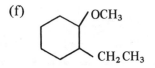

(g)

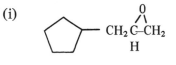

(h)   CH₂ –0–CH₃

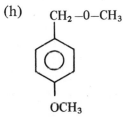

(i)

(j)   I

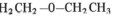

$CH_2CH_2-O-CH_2CH_3$

(k)

(Hint, K requires a glycol)

CHAPTER VIII

CARBOXYLIC ACIDS

PROCEDURES FOR SYNTHESIZING CARBOXYLIC ACIDS

VIII-1 Carbonation of a Grignard reagent

$$RMgX \xrightarrow{\quad CO_2 \quad} \underset{\underset{O}{\parallel}}{RCOMgX} \xrightarrow{\quad HCl \quad} \underset{\underset{O}{\parallel}}{RCOH}$$

There are two restrictions concerning the use of this method:

(1) the Grignard reagent can be made
(2) there are no other groups on the Grignard molecule which will react with it before the Grignard reagent has a chance to be carbonated.

VIII-2 Hydrolysis of a nitrile

(a) Basic conditions: used when there are acid sensitive groups in the molecule.

$$RC\equiv N \xrightarrow{\quad NaOH, \Delta \quad} \underset{\underset{O}{\parallel}}{RCO^- Na^+} \xrightarrow{\quad HCl \quad} \underset{\underset{O}{\parallel}}{RCOH}$$

$$\underset{\underset{OH}{|}}{CH_3\overset{\overset{CH_3}{|}}{C}CH_2CH_2C\equiv N} \xrightarrow[\text{(2) HCl}]{\text{(1) NaOH, }\Delta} \underset{\underset{OH}{|}}{CH_3\overset{\overset{CH_3}{|}}{C}CH_2CH_2\overset{\overset{O}{\parallel}}{C}OH}$$

(b) Acidic conditions: used when there are base sensitive groups present in the molecule.

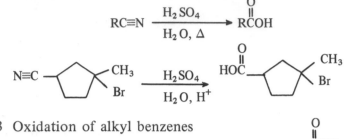

$$RC{\equiv}N \xrightarrow[\text{H}_2\text{O, }\Delta]{\text{H}_2\text{SO}_4} \overset{\overset{\displaystyle O}{\|}}{R}C\text{OH}$$

VIII-3 Oxidation of alkyl benzenes

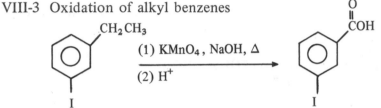

$Na_2Cr_2O_7$ and H_2SO_4 may be used in place of $KMnO_4$ and ⁻OH. This reaction will not proceed if the carbon attached to the ring is tertiary.

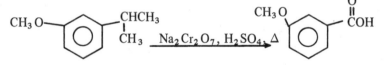

VIII-4 Oxidation of primary alcohols

$$RCH_2OH \xrightarrow{\text{Na}_2\text{Cr}_2\text{O}_7,\ \text{H}_2\text{SO}_4,\ \Delta} R\overset{\overset{\displaystyle O}{\|}}{C}OH$$

$$CH_3CH_2CH_2OH \xrightarrow{\text{Na}_2\text{Cr}_2\text{O}_7,\ \text{H}_2\text{SO}_4,\ \Delta} CH_3CH_2\overset{\overset{\displaystyle O}{\|}}{C}OH$$

$KMnO_4$ and $NaOH$ can be used in place of $Na_2Cr_2O_7$ and H_2SO_4 if there are acid sensitive groups present in the molecule.

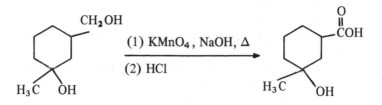

$$\text{(1) KMnO}_4\text{, NaOH, }\Delta$$

$$\text{(2) HCl}$$

WORKED SYNTHESES

Using any monosubstituted cyclic organic molecule, any mono-substituted acyclic organic molecule containing no more than four carbons and any inorganic reagent prepare the following.

VIII-a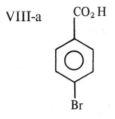

Let us ask some questions concerning VIII-a.

(1) What kind of molecule is VIII-a?
    Answer: VIII-a is an aromatic carboxylic acid with a Br in the *para* position (4-bromobenzoic acid).

(2) What procedures are available for preparing aromatic carboxylic acids?
    Answer: VIII-1, VIII-2, VIII-3, VIII-4

    Procedure VIII-1 would necessitate the formation of a Grignard reagent from a *para* dihalide. This is not normally a good reaction if the unreacted halogen is Br or I (polymer formation occurs).

    Procedure VIII-2 requires the presence of a nitrile group ($-C\equiv N$). Since we have not covered the methods of placing a $-C\equiv N$ on an aromatic ring, we will bypass this procedure.

    Procedure VIII-3 requires the presence of an alkyl group such as $-CH_3$ or $-C_2H_5$ on the ring. This is possible.

Procedure VIII-4 requires an alcohol which would normally be formed from a Grignard reagent. This could lead to the same complications as stated above for procedure VIII-1.

From the above, it is concluded that procedure VIII-3 is the method of choice.

(3) Do we have to add carbon atoms?
Answer: Since we are allowed to start with a monosubstituted aromatic ring, we have the option of starting with

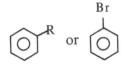

If we start with the former, we do not have to add carbons.

(4) What is the structure of the alkyl benzene needed so that via procedure VIII-3, VIII-a will be produced?
Answer: If R is –CH₃, structure **A**.

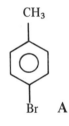

(5) How can **A** be produced?
Answer: If we start with toluene, procedure V-4 will allow us to place the Br on the ring. Toluene is a permitted starting material.

Thus the synthesis of VIII-a, working backwards in a stepwise manner becomes:

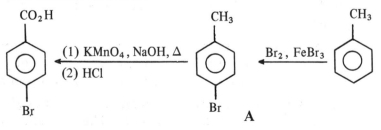

VIII-b

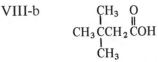

Let us ask some questions concerning VIII-b.

(1) What kind of molecule is VIII-b?
Answer: VIII-b is an aliphatic carboxylic acid (3,3-dimethyl-butanoic acid).

(2) Do we have to add carbon atoms?
Answer: Yes, since the maximum number of carbons we are permitted to start with in preparing an acyclic molecule is four and VIII-b contains six carbons.

(3) What procedures are available for the preparation of aliphatic carboxylic acids?
Answer: Procedures VIII-1, VIII-2 and VIII-4.

Procedure VIII-1 requires the Grignard reagent A.

$$CH_3\underset{\underset{CH_3}{|}}{\overset{\overset{CH_3}{|}}{C}}CH_2MgBr \qquad A$$

Procedure VIII-2 requires the nitrile B.

$$CH_3\underset{\underset{CH_3}{|}}{\overset{\overset{CH_3}{|}}{C}}CH_2C{\equiv}N \qquad B$$

Procedure VIII-4 requires the alcohol C.

$$CH_3\underset{\underset{CH_3}{|}}{\overset{\overset{CH_3}{|}}{C}}CH_2CH_2OH \qquad C$$

Let us select C and use procedure VIII-4. As an exercise, the student may use A or B as the precursor of VIII-b.

(4) How can C be prepared?
Answer: Since C is a primary alcohol with the structure $-CH_2CH_2OH$, it can be prepared from the Grignard agent D and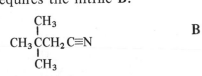

$$
\begin{array}{c}
\text{CH}_3 \\
| \\
\text{CH}_3\,\text{C}-\text{MgBr} \qquad\qquad \textbf{D} \\
| \\
\text{CH}_3
\end{array}
$$

Since **D** contains only four carbons, the synthesis of VIII-b becomes:

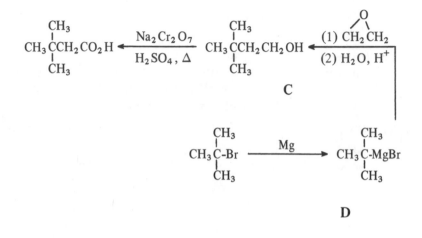

VIII-c

$$
\begin{array}{c}
\text{CH}_3 \\
| \\
\text{CH}_3\,\text{CHCH}_2\,\text{CHCO}_2\,\text{H} \\
| \\
\text{CH}_3
\end{array}
$$

Let us ask some questions concerning VIII-c.

(1) What kind of molecule is VIII-c?
Answer: VIII-c is an aliphatic carboxylic acid (2,4-dimethyl-pentanoic acid).

(2) Do we have to add carbon atoms?
Answer: Yes, because VIII-c contains seven carbons while we are permitted to start with acyclic molecules containing no more than four carbons.

(3) What procedures are available for preparing aliphatic acyclic carboxylic acids?
Answer: Procedures VIII-1, VIII-2 and VIII-4.

Let us select procedure VIII-1. (The student may select procedures VIII-2 and VIII-4 as an exercise.) Procedure

VIII-1 requires the Grignard reagent made from the alkyl halide **A**.

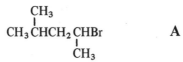

$$\underset{\underset{CH_3}{|}}{\overset{\overset{CH_3}{|}}{CH_3 CHCH_2 CHBr}} \qquad A$$

(4) How can **A** be prepared?
Answer: **A** can be prepared from the olefin **B** via procedure III-1.

$$\underset{\underset{H}{|}}{\overset{\overset{CH_3}{|}}{CH_3 CHCH_2 C}}=CH_2 \qquad B$$

(5) How can **B** be prepared?
Answer: Keeping in mind that we have to add carbon atoms (**B** contains six carbons) and that the Grignard reaction is an excellent procedure for adding carbons, let us dehydrate the alcohol **C**.

$$\underset{\underset{H}{|}}{\overset{\overset{CH_3}{|}}{CH_3 - C - CH_2CH_2CH_2OH}} \qquad C$$

(6) How can **C** be prepared?
Answer: **C** is a primary alcohol containing the structure $-CH_2 CH_2 OH$, thus it can be prepared by treatment of the Grignard reagent **D** with $CH_2 CH_2$.
$$\underset{O}{\overset{}{\diagdown \diagup}}$$

$$\underset{\underset{CH_3}{|}}{\overset{\overset{CH_3}{|}}{CH_3 CHCH_2 MgBr}} \qquad D$$

Noting that **D** contains four carbons the synthesis of VIII-c becomes

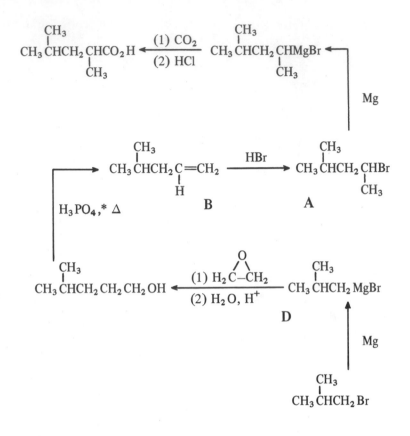

---

*C is a primary alcohol and is quite prone to undergo carbon skeleton rearrangement when treated with hot acid. To obviate the carbon skeleton rearrangement and effect the dehydration so that **B** is the major product, it is suggested that the acetate ester of **C** be formed and pyrolyzed as shown below. The acetate ester pyrolysis is known to yield olefins in which there is no carbon skeleton rearrangement.

$$CH_3 CHCH_2 CH_2 CH_2 OH \xrightarrow{(CH_3 C)_2 O} CH_3 CHCH_2 CH_2 CH_2 \xrightarrow{300°C} B$$

VIII-d

Let us ask some questions concerning VIII-d.

(1) Answer: VIII-d is a cyclic aliphatic carboxylic acid (2-methyl-cyclopentane carboxylic acid).

(2) Do we have to add carbon atoms?
Answer: Yes, since VIII-d is a disubstituted acyclic molecule and each substituent contains carbon. We are permitted to start with only a monosubstituted organic molecule.

(3) What procedures are available for producing an aliphatic cyclic carboxylic acid?
Answer: Procedures VIII-1, VIII-2 and VIII-4.

Let us use procedure VIII-2. The use of procedures VIII-1 and VIII-4 will be left to the student as an exercise.

The use of procedure VIII-2 requires the nitrile **A**.

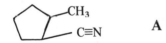 **A**

(4) How can **A** be prepared?
Answer: A nitrile may be placed on an aliphatic carbon by an $S_N 2$ type reaction. Thus, treatment of **B** with KCN would yield **A**.

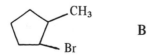 **B**

(5) How can **B** be prepared?
Answer: **B**   can be prepared by treatment of olefin **C** with HBr and peroxide via procedure III-2.

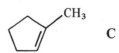 **C**

(6) How can **C** be prepared?
Answer: Mindful that we must place the CH_3 group on the ring, a Grignard synthesis comes to mind. Thus the dehydration of alcohol **D** via procedure II-2 is the method of choice.

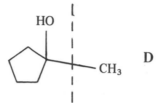

(7) How can **D** be prepared?
Answer: **D** is a tertiary alcohol, thus if we cleave the CH_3 ring linkage as illustrated above, we can see that the required ketone is **E**, a permitted starting material.

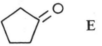

Thus the synthesis of VIII-d is

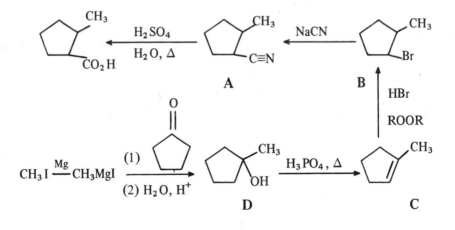

UNWORKED SYNTHESES

Using any monosubstituted cyclic molecule, any monosubstituted acyclic organic molecule containing no more than four carbons and any inorganic reagents, prepare the following compounds.

(e) $CH_3CH_2CH_2CH_2CH_2CH_2CO_2H$ (f)

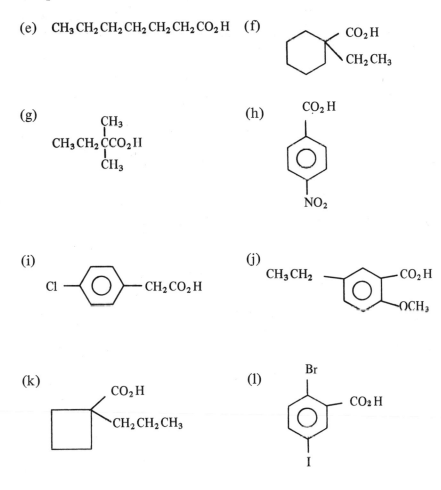

(g)

$CH_3CH_2\overset{\overset{\displaystyle CH_3}{|}}{\underset{\underset{\displaystyle CH_3}{|}}{C}}CO_2H$

(h)

(i)

(j)

(k)

(l)

CHAPTER IX

CARBOXYLIC ACID DERIVATIVES

A. ACID HALIDES $R-\overset{\overset{\displaystyle O}{\|}}{C}-X$ (X = Cl, Br)

PROCEDURES FOR SYNTHESIZING ACID HALIDES

IX-1 Acid chlorides: reaction of a carboxylic acid + PCl_3, PCl_5 or $SOCl_2$

$$\underset{\text{RCOH}}{\overset{\overset{\displaystyle O}{\|}}{}} \xrightarrow{PCl_3} \underset{\text{RC-Cl}}{\overset{\overset{\displaystyle O}{\|}}{}}$$

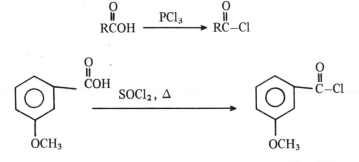

IX-2 Acid bromides: reaction of a carboxylic acid + PBr_3 or PBr_5

$$\underset{\text{RCOH}}{\overset{\overset{\displaystyle O}{\|}}{}} \xrightarrow{PBr_3} \underset{\text{RC-Br}}{\overset{\overset{\displaystyle O}{\|}}{}}$$

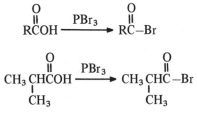

117

B. ESTERS R–C̈–O–R′

PROCEDURES FOR SYNTHESIZING ESTERS

IX-3 Fischer esterification

$$\underset{\text{RC}-\text{OH}}{\overset{\text{O}}{\|}} + \text{R}′\text{OH} \;\underset{}{\overset{\text{H}^+}{\rightleftharpoons}}\; \underset{\text{RC}-\text{OR}′}{\overset{\text{O}}{\|}} + \text{H}_2\text{O}$$

$$\underset{\text{CH}_3\text{CHCH}_2\text{CO}_2\text{H}}{\overset{\text{CH}_3}{|}} + \text{CH}_3\text{OH} \;\overset{\text{H}^+}{\longrightarrow}\; \underset{\text{CH}_3\text{CHCH}_2\text{COCH}_3}{\overset{\text{CH}_3\quad\text{O}}{|\qquad\|}}$$

Since this reaction is equilibrium controlled, a sufficient yield of the ester can be quaranteed only if there is a large excess of the alcohol present. Thus, this reaction is normally used if the alcohol is inexpensive and is in abundant supply.

IX-4 Reaction of an acid chloride and an alcohol.

$$\underset{\text{RCCl}}{\overset{\text{O}}{\|}} + \text{R}′\text{OH} \;\overset{\text{base}}{\longrightarrow}\; \underset{\text{RCOR}′}{\overset{\text{O}}{\|}}$$

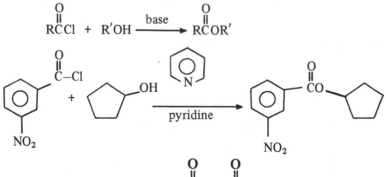

C. ACID ANHYDRIDES R–C̈–O–C̈–R

PROCEDURES FOR SYNTHESIZING ACID ANHYDRIDES

IX-5 Symmetrical anhydride: reaction of a carboxylic acid and phosphorous pentoxide (P_2O_5) or acetic anhydride [$(CH_3\overset{O}{\overset{\|}{C}})_2O$ or Ac_2O].

$$RCO_2H \xrightarrow{P_2O_5} \overset{\overset{O}{\|}}{R}C\overset{\overset{O}{\|}}{O}CR$$

$$CH_3CH_2CH_2CO_2H \xrightarrow{Ac_2O} CH_3CH_2CH_2\overset{\overset{O}{\|}}{C}O\overset{\overset{O}{\|}}{C}CH_2CH_2CH_3$$

IX-6 Unsymmetrical anhydrides: reaction of an acid chloride and a salt of a carboxylic acid.

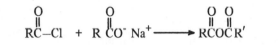

$$R\overset{\overset{O}{\|}}{C}-Cl \ + \ R\overset{\overset{O}{\|}}{C}O^- Na^+ \longrightarrow R\overset{\overset{O}{\|}}{C}O\overset{\overset{O}{\|}}{C}R'$$

$$CH_3CH_2\overset{\overset{O}{\|}}{C}-Cl \ + \ CH_3\underset{\underset{CH_3}{|}}{CH}\overset{\overset{O}{\|}}{C}O^- Na^+ \longrightarrow CH_3CH_2\overset{\overset{O}{\|}}{C}O\overset{\overset{O}{\|}}{C}\underset{\underset{CH_3}{|}}{CH}CH_3$$

D. AMIDES $R-\overset{\overset{O}{\|}}{C}NR_2 \, (H)$

PROCEDURES FOR SYNTHESIZING AMIDES

IX-7 Reaction of a carboxylic acid chloride with an amine.

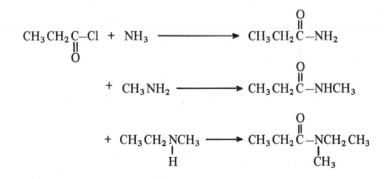

$$CH_3CH_2\underset{\underset{O}{\|}}{C}-Cl \ + \ NH_3 \longrightarrow CH_3CH_2\overset{\overset{O}{\|}}{C}-NH_2$$

$$+ \ CH_3NH_2 \longrightarrow CH_3CH_2\overset{\overset{O}{\|}}{C}-NHCH_3$$

$$+ \ CH_3CH_2\underset{\underset{H}{|}}{N}CH_3 \longrightarrow CH_3CH_2\overset{\overset{O}{\|}}{C}-\underset{\underset{CH_3}{|}}{N}CH_2CH_3$$

The amine used in the reaction must have at least one H attached to the N.

WORKED SYNTHESES

Using any monosubstituted cyclic organic molecule, any mono-substituted acyclic organic molecule containing no more than four carbons, any amine and any inorganic reagents, prepare the following compounds.

IX-a

$$CH_3CH_2O-\overset{\overset{O}{\|}}{C}-\langle\!\langle\ \rangle\!\rangle-OCH_3$$

Let us ask some questions concerning IX-a.

(1) What kind of molecule is IX-a?
Answer: IX-a is an aromatic ester (ethyl 4-methoxybenzoate).

(2) Do we have to add carbon atoms?
Answer: Yes, because IX-a is a disubstituted cyclic molecule with each substituent containing carbon atoms.

(3) How can IX-a be broken up so that the acid and alcohol portions of the ester can be discerned?
Answer: IX-a may be broken up by cleaving the bond as illustrated below. This gives us the alcohol **A** and the car-boxylic acid **B** which compose IX-a.

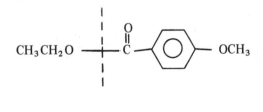

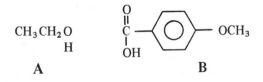

A   B

(4) What procedures are available for preparing esters?
Answer: Procedures IX-3 and IX-4. The Fischer preparation may be used provided the alcohol is abundant. **A** (ethyl

alcohol) is inexpensive, thus we shall use procedure IX-3. Procedure IX-4 may also be used to prepare IX-a; this will be left as an exercise for the student.

(5) How can **B** be prepared?
Answer: Since **B** is an aromatic carboxylic acid, procedures VIII-1, VIII-2 and VIII-4 will suffice. Let us use procedure VIII-1. This will necessitate the Grignard reagent formed from the halide **C**.

Br —⟨O⟩— OCH$_3$ **C**

(6) How can **C** be prepared?
Answer: By reacting anisole with Br$_2$ via procedure V-4. Anisole is a permitted starting material.

The synthesis of IX-a, working backwards in a stepwise fashion, now becomes:

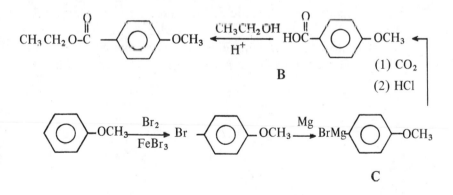

IX-b
$$CH_3CH_2\underset{\underset{H}{\overset{}{|}}}{\underset{\underset{CH_3}{|}}{CH}}-\overset{\overset{O}{\|}}{C}-NCH_2CH_2CH_2CH_3$$

Let us ask some questions concerning IX-b.

(1) What kind of molecule is IX-b?
Answer: IX-b is an aliphatic secondary amide (two carbons attached to N) (N-butyl-2-methylbutanamide).

(2) Do we have to add carbon atoms?
Answer: Yes, because IX-b contains nine carbons.

(3) How can IX-b be broken up so that the carboxylic acid and amine needed for the synthesis of IX-b can be determined?
Answer: IX-b can be broken up by cleaving the bond
$$\overset{O}{}$$
between the $-C$ and N as illustrated below. This gives us the required carboxylic acid portion **A** and the amine portion **B**.

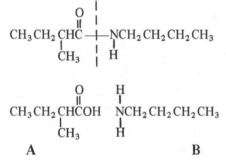

$$\underset{\underset{\underset{H}{|}}{CH_3}}{\underset{|}{CH_3CH_2CHCOH}} \quad \underset{\underset{H}{|}}{\overset{H}{\overset{|}{N}}CH_2CH_2CH_2CH_3}$$

A   **B**

(4) What procedure is available for the preparation of amides?
Answer: Procedure IX-7, which requires a carboxylic acid chloride cleaved from the acid portion of the amide and an amine. Thus, for IX-b we need the acid chloride **C** and the amine **B**.

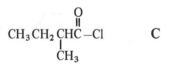

$$\underset{\underset{CH_3}{|}}{CH_3CH_2CHC-Cl} \qquad \textbf{C}$$

(5) How can **C** be prepared?
Answer: Since **C** is an acid chloride it can be made by reacting **A** with $SOCl_2$ via procedure IX-1.

(6) How can **A** be prepared?
Answer: Since **A** is an aliphatic carboxylic acid, it can be synthesized from **D** via procedure VIII-2a.

$$\underset{\underset{CH_3}{|}}{CH_3CH_2CHC\equiv N} \qquad \textbf{D}$$

We will assume the substituent (−C≡N) even though it contains a carbon will not count against the number of carbons in the organic molecule. Thus, we can start with **D** and the synthesis of IX-b becomes:

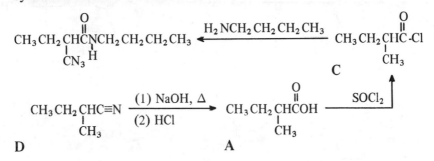

$$CH_3 CH_2 \overset{O}{\overset{\|}{C}} H \overset{\|}{N} CH_2 CH_2 CH_2 CH_3 \xleftarrow{\quad H_2 NCH_2 CH_2 CH_2 CH_3 \quad} CH_3 CH_2 \overset{O}{\overset{\|}{C}} HC\text{-}Cl$$

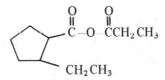

$$CH_3 CH_2 \underset{\underset{CH_3}{|}}{C} HC\equiv N \xrightarrow[\text{(2) HCl}]{\text{(1) NaOH, } \Delta} CH_3 CH_2 \underset{\underset{CH_3}{|}}{C} H\overset{O}{\overset{\|}{C}} OH \xleftarrow{\quad SOCl_2 \quad}$$

D   **A**

IX-c

Let us ask some questions concerning IX-c.

(1) What kind of molecule is IX-c?
Answer: IX-c is an unsymmetrical aliphatic anhydride (propanoic 2-ethyl-cyclopentanecarboxylic acid anhydride).

(2) Do we have to add carbon atoms?
Answer: Yes, because IX-c is a disubstituted cyclic molecule and each substituent contains carbon atoms.

(3) What procedure is available for preparing unsymmetrical anhydrides?
Answer: Procedure IX-6 which requires an acid chloride and a salt of a carboxylic acid.

(4) How can IX-c be broken up so that the structures of the required acid chloride and acid salt can be determined?
Answer: To determine the structures of the required starting materials, the single bond between the $-\overset{O}{\overset{\|}{C}}-$ and the −O− must be cleaved. Since there are two $\overset{O}{\overset{\|}{C}}$−O linkages the molecule may be cleaved in two ways as shown below.

Cleavage 1   Cleavage 2

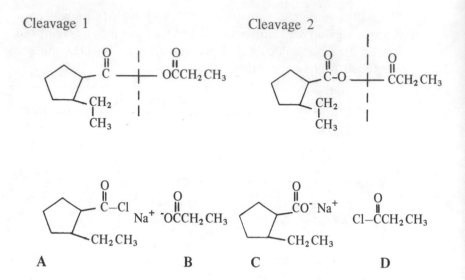

A   B   C   D

Cleavage 1 requires the use of **A** and **B** while cleavage 2 requires **C** and **D**. We shall use **A** and **B**. The student may use **C** and **D** as an exercise. Note that we can start with **B** so we must prepare only **A**.

(5) How can **A** be prepared?
Answer: **A** can be prepared from the corresponding acid **E** upon treatment with PCl_3 via procedure IX-1.

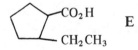

(6) How can **E** be prepared?
Answer: **E** being a carboxylic acid can be prepared via procedures VIII-1, VIII-2 and VIII-4. Let us select procedure VIII-1, the carbonation of the Grignard reagent derived from the alkyl bromide **F**.

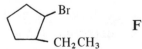

(7) How can **F** be prepared?
Answer: **F** can be prepared by treating olefin **G** with HBr and peroxide via procedure III-2.

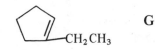

G

(8) How can **G** be prepared?
Answer: **G** can be prepared by dehydration of the alcohol **H** via procedure II-2.

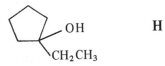

H

(9) How can **H** be prepared?
Answer: **H** being a tertiary alcohol can be prepared by the reaction of the ketone **I** with CH_3CH_2MgBr.

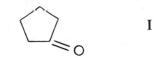

I

The synthesis of IX-c now becomes:

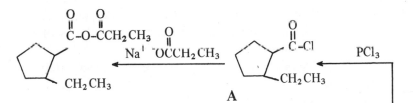

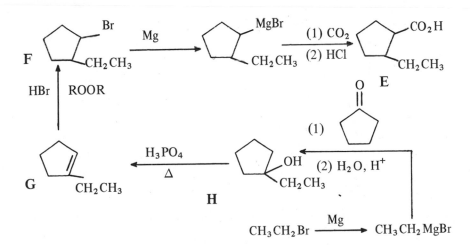

IX-d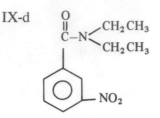

Let us ask some questions concerning IX-d.

(1)   What kind of molecule is IX-d?
Answer: IX-d is an aromatic tertiary amide (three carbons attached to the N) (N,N-diethyl-3-nitrobenzamide).

(2)   Do we have to add carbon atoms?
Answer: IX-d is a disubstituted cyclic molecule. Since only one of the substituents contains carbon, we may or may not have to add carbon atoms.

(3)   What procedure is available for the preparation of a tertiary amide?
Answer: Procedure IX-7.

(4)   How can IX-d be broken up so that the structures of the acid chloride and amine required for the synthesis of IX-d can be deduced?
Answer: The structures of the starting materials required for procedure IX-7 can be deduced by cleaving the bond between the $-\overset{\text{O}}{\overset{\|}{\text{C}}}-$ and the $-\overset{|}{\text{N}}-$ atom as illustrated below.

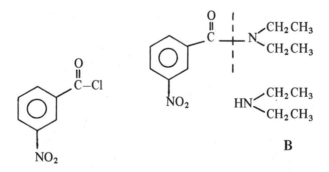

We can start with **B**, but **A** must be synthesized.

(5) How can **A** be prepared?

Answer: **A** can be prepared by treating carboxylic acid **C** with $SOCl_2$ via procedure IX-1.

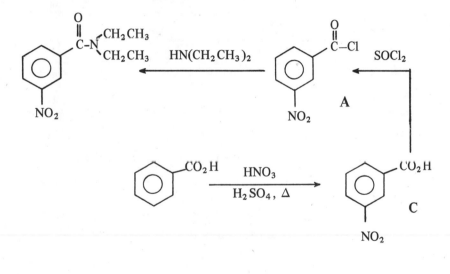

(6) How can **C** be prepared?

Answer: Since the $-CO_2H$ group is a *meta* director, we can nitrate benzoic acid via procedure V-6.

Thus the synthesis of IX-d becomes:

IX-e

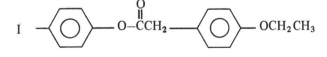

Let us ask some questions concerning IX-e.

(1) What kind of molecule is IX-e?

Answer: IX-e is an ester containing two aromatic rings (4'-iodophenyl 4-ethoxyphenylacetate).

(2) Do we have to add carbon atoms?
Answer: Yes, because one of the rings contains two substituents each of which contain carbon atoms.

(3) What procedures are available for preparing esters?
Answer: Procedures IX-3 and IX-4. To see which procedure
we should use, we first have to break the molecule up into
its component parts.

(4) How can IX-e be broken up so that the structures of the
acid and alcohol which make up IX-e can be deduced?
Answer: To ascertain the structures of the acid and alcohol

that compose IX-e, the $\overset{O}{\overset{\|}{C}}$–O bond must be cleaved as illus-
trated below.

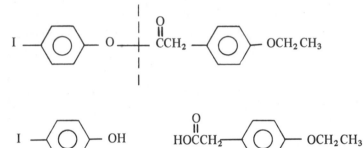

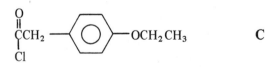

A   B

Since the alcohol (phenol) **A** is not inexpensive we shall use
procedure IX-4.which requires the acid chloride **C**.

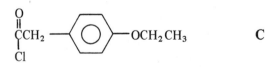

C

We shall now set about synthesizing the starting materials
A and **C**.

(5) How can **A** be prepared?
Answer: Procedure V-5, the iodination of aromatic rings
can be modified if the aromatic ring is a phenol. We merely
have to treat the phenol with I_2.

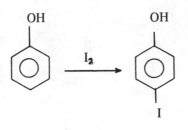

(6) How can **C** be prepared?
Answer: **C** can be prepared by treating **B** with $SOCl_2$ via procedure IX-1.

(7) How can **B** be prepared?
Answer: **B** being a carboxylic acid can be prepared by hydrolyzing the nitrile **D** via procedure VIII-2b.

$$N{\equiv}CCH_2{-}\overset{}{\bigcirc}{-}OCH_2CH_3 \qquad \textbf{D}$$

(8) How can **D** be prepared?
Answer: **D** can be prepared by treating the benzyl chloride **E** with KCN as shown in problem VIII-d.

$$\underset{\underset{Cl}{|}}{CH_2}{-}\overset{}{\bigcirc}{-}OCH_2CH_3 \qquad \textbf{E}$$

(9) How can **E** be prepared?
Answer: Since **E** is a benzyl halide it can easily be synthesized by reacting the corresponding benzyl alcohol with HCl. Another method of synthesizing benzyl halides involves treating an activated aromatic ring with HCl, CH_2O and $ZnCl_2$. The reaction is called chloromethylation and is illustrated below.

$$R{-}\overset{}{\bigcirc} \xrightarrow{H_2C{=}O,\ HCl,\ ZnCl_2} R{-}\overset{}{\bigcirc}{-}CH_2Cl$$

To synthesize **E** by this procedure would require the molecule phenetole, a permitted starting material. Let us now synthesize IX-e.

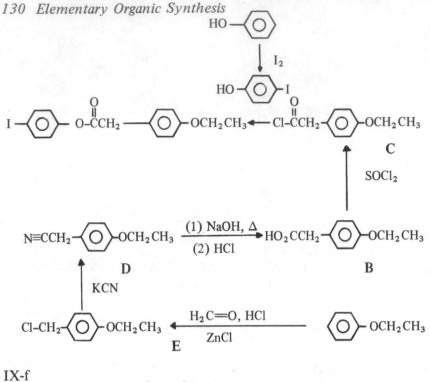

IX-f

$$CH_3CH_2\underset{\underset{Cl}{|}}{\overset{\overset{O}{\parallel}}{C}}HC-O-\underset{\underset{CH_3}{|}}{\overset{\overset{O}{\parallel}}{C}}CHCH_2-\left\langle\bigcirc\right\rangle-C(CH_3)_3$$

As in the case of IX-e this synthesis will require the use of reactions covered in the lecture, but not specifically stated in the preparation procedures in this book. The reactions needed will be detailed during the course of this preparation. Let us ask some questions concerning IX-f.

(1) What kind of molecule is IX-f?
Answer: IX-f is an unsymmetrical anhydride [2-chlorobutanoic 2-methyl-3-(p-t-butylphenyl)propanoic anhydride].

(2) Do we have to add carbon atoms?
Answer: Since IX-f contains a disubstituted ring with each substituent containing carbon atoms, yes, we do have to add carbon atoms.

(3) What procedure is available for preparing unsymmetrical anhydrides?
Answer: Procedure IX-6 requiring a carboxylic acid chloride and a salt of a carboxylic acid.

(4) How can IX-f be broken up so that the structure of the
required acid chloride and acid salt can be deduced?
Answer: The structures of the required starting materials
can be found by cleaving the $\overset{O}{\overset{\|}{C}}$–O bond. Since the anhy-
dride possesses two such bonds the molecule can be cleaved
in two ways as illustrated below.

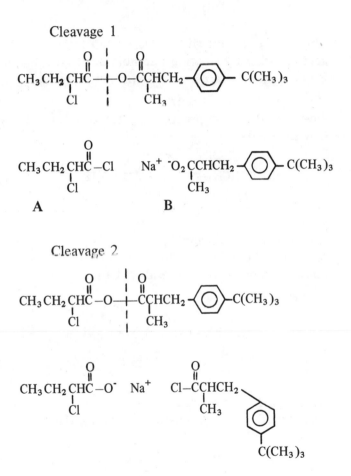

Let us use cleavage 1 requiring **A** and **B**. Synthesizing IX-f
via the starting materials from cleavage 2 will be left to the
student as an exercise.

(5) How can **A** be prepared?
Answer: **A** is an acid chloride containing a chloride on the α-carbon. This can be prepared by using the Hell-Volhard-Zelinsky reaction shown below.

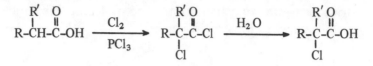

Thus, if we do not add H_2O, we can isolate the α-chlorinated acid chloride by treating butanoic acid with Cl_2 and PCl_3. Butanoic acid is a permitted starting material.

(6) How can **B** be prepared?
Answer: **B** can be prepared by treating the carboxylic acid **C** with NaOH.

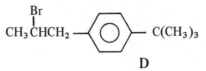

(7) How can **C** be prepared?
Answer: **C** can be prepared by carbonating the Grignard reagent prepared from the alkyl halide **D** via procedure VIII-1.

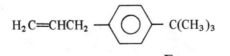

(8) How can **D** be prepared?
Answer: **D** can be prepared by adding HBr to olefin **E** via procedure III-1.

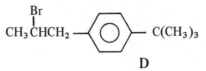

(9) How can **E** be prepared?
Answer: The CH_2=CH–CH_2– group is the allyl group.
Allyl halides couple with Grignard reagents very readily as
shown below:

$$H_2C=CHCH_2-Br \quad + \quad RMgX \longrightarrow RCH_2C=CH_2$$
$$H$$

F

Thus, if we add allyl bromide **F** to the Grignard reagent
made from the aryl halide **G**, we can synthesize **E**.

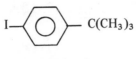

G

(10) How can **G** be prepared?
Answer: **G** can be prepared by iodinating *t*-butylbenzene
via procedure V-5. Thus, the synthesis of IX-f becomes

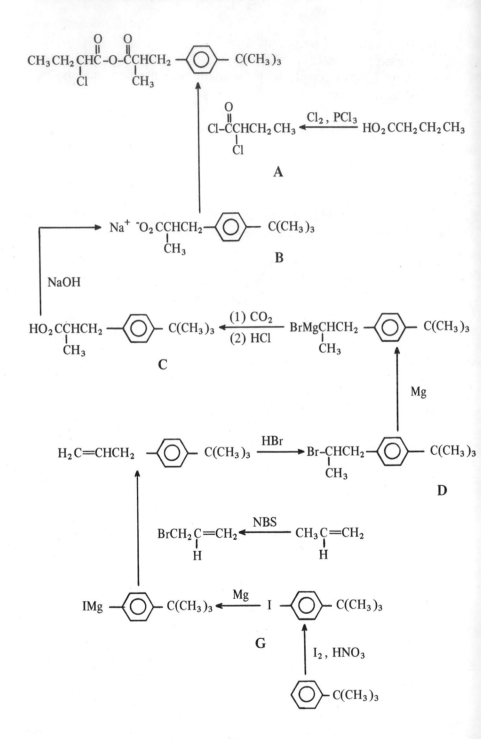

UNWORKED SYNTHESES

Synthesize the following using the instructions at the beginning of the worked syntheses problems section.

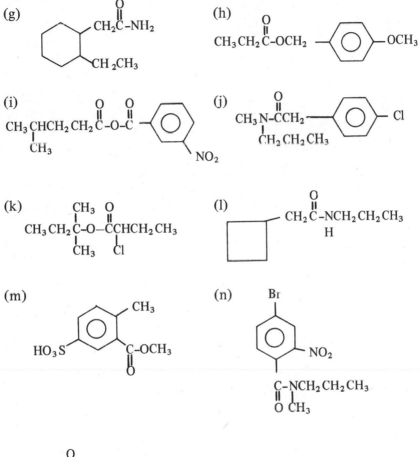

(g)

(h)

(i)

(j)

(k)

(l)

(m)

(n)

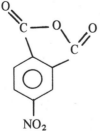

(o)

ALDEHYDES AND KETONES

PROCEDURES FOR SYNTHESIZING ALDEHYDES

X-1 Stephen's reduction

$$R{-}C{\equiv}N \xrightarrow[\text{(2) } H_2O]{\text{(1) } SnCl_2,\ HCl} R{-}\overset{\overset{\displaystyle O}{\|}}{C}H$$

This reaction works best for aromatic aldehydes and for aliphatic chains up to seven carbons.

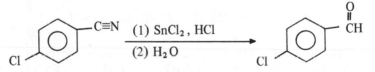

X-2 LiAlH[OC(CH$_3$)$_3$]$_3$ reduction

The conversion of an acid chloride into the corresponding aldehyde requires the use of the reagent lithium aluminum tri-t-butoxy hydride (LiAlH[OC(CH$_3$)$_3$]$_3$) or LiAlH(O-t-Bu)$_3$

This reaction works on both aliphatic and aromatic acid chlorides.

$$R\overset{\overset{\displaystyle O}{\|}}{C}{-}Cl \xrightarrow{\text{LiAlH(O-}t\text{-Bu)}_3} R{-}\overset{\overset{\displaystyle O}{\|}}{C}H$$

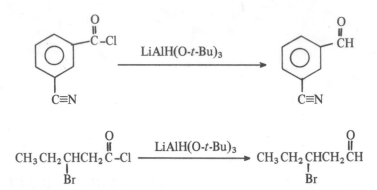

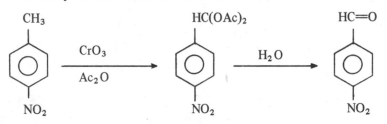

X-3 Chromium trioxide (CrO_3) and acetic anhydride (Ac_2O) oxidation. This reaction is good only for the formation of aromatic aldehydes. This reaction requires a $-CH_3$ attached directly to the aromatic nucleus.

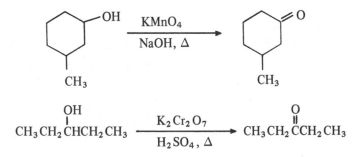

PROCEDURES FOR SYNTHESIZING KETONES

X-4 Oxidation of a secondary alcohol requires the use of a strong oxidizing agent, either $KMnO_4$ and OH or $K_2Cr_2O_7$ and H_2SO_4 (the latter is more frequently used).

Organic Synthesis

required, AlCl₃ may be replaced by SnCl₄ or

permits the formation of aromatic rings con-
atic straight chain substituents. This is accom-
educing the ketone formed via the Clemmensen
the Wolff-Kischner reduction.

uction

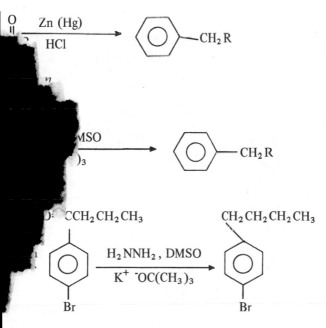

is the solvent dimethyl sulfoxide.

ESES

substituted cyclic organic molecule, any mono-
organic molecule containing no more than five
norganic reagents, prepare the following.

O
‖
HCH

H₃

X-5 The reaction of a cadmium Grignard reage
an acid chloride. This reaction requires th
a cadmium Grignard which can be formed

$$RMgX + CdCl_2 \rightarrow R_2Cd + $$

The acid chloride may be aliphatic or arom

$$CH_3\underset{\underset{CH_3}{|}}{C}HMgBr + CdCl_2 \longrightarrow (CH_3\underset{\underset{CH_3}{|}}{C}H-)_2Cd$$

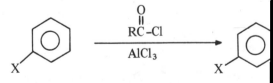

X-6 The Friedel-Crafts Acylation reaction

Several points concerning this reaction wh
remembered:

(1) If $X = NO_2$, CN, CHO, $\overset{\overset{O}{\|}}{C}$-R, CO_2R, SO_3
will not go. (However, if a molecule poss
electron donating group in addition to on
mentioned above, the reaction will go wit
donor group dictating the position of sub

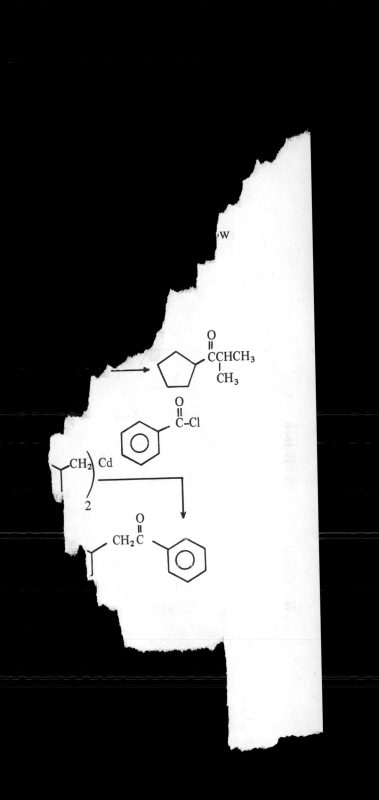

(3)

p
redu

Clemmensen

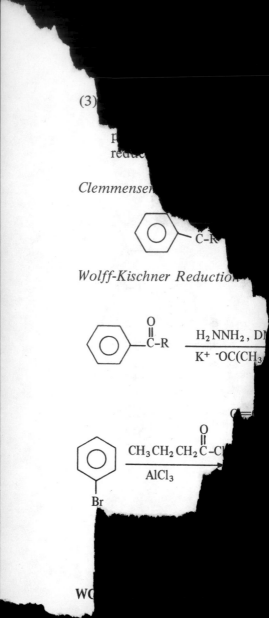

C-R

Wolff-Kischner Reduction

$$\text{C-R} \quad \frac{H_2NNH_2, \ DM}{K^+ \ ^-OC(CH_3}$$

O=

$$\text{Br} \quad \frac{CH_3CH_2CH_2\overset{O}{\overset{\|}{C}}-C}{AlCl_3}$$

WO

Let us ask some questions concerning X-a.

(1) What kind of molecule is X-a?
Answer: X-a is an acyclic aliphatic aldehyde (2-methyl-pentanal).

(2) Do we have to add carbon atoms?
Answer: Since the molecule contains six carbons and we are permitted to start with a molecule containing no more than five carbons, we must add carbon atoms.

(3) What procedures are available for preparing an aliphatic aldehyde?
Answer: Procedures X-1 and X-2. Let us use procedure X-2. (The preparation of X-a via procedure X-1 will be left as an exercise to the student). Procedure X-2 will require the use of the acid chloride **A**.

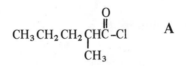

$$CH_3CH_2CH_2\underset{\underset{CH_3}{|}}{CH}\overset{\overset{O}{\|}}{C}-Cl \qquad \textbf{A}$$

(4) How can **A** be prepared?
Answer: **A** being an acid chloride can be prepared from the carboxylic acid **B** using procedure IX-1.

$$CH_3CH_2CH_2\underset{\underset{CH_3}{|}}{CH}CO_2H \qquad \textbf{B}$$

Note: **B** contains six carbons and must be prepared.

(5) How can **B** be prepared?
Answer: **B** being a carboxylic acid can be prepared by treating the Grignard reagent from **C** via procedure VIII-1.

$$CH_3CH_2CH_2\underset{\underset{CH_3}{|}}{CH}-Br \qquad \textbf{C}$$

Note: We can start with **C** since it contains only five carbons and is monosubstituted.

The synthesis of X-a working backwards in a stepwise fashion now becomes:

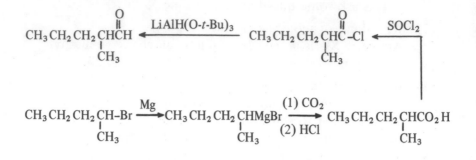

X-b

$$\underset{CH_3CH_2}{}\overset{}{\bigcirc}-CH_2\overset{O}{\overset{\|}{C}}\underset{\underset{CH_3}{|}}{C}HCH_3$$

Let us ask some questions concerning X-b.

(1) What kind of molecule is X-b?
Answer: X-b is an aliphatic ketone containing a *para* substituted aromatic ring [3-methyl-1-(p-ethylphenyl)-2-butanone].

(2) Do we have to add carbon atoms?
Answer: Yes, because X-b is a disubstituted ring and each substituent contains carbon atoms.

(3) What procedures are available for preparing aliphatic ketones?
Answer: Procedures X-4 and X-5. Let us use procedure X-5. (The preparation of X-b via procedure X-4 will be left to the student as an exercise.)

(4) How can X-b be broken up so that the required starting materials needed for procedure X-5 can be discerned?
Answer: X-b being a ketone can be broken up by cleaving either of the C–C bonds. Since there are two such bonds, the molecule can be cleaved in two ways as illustrated below.

Cleavage 1

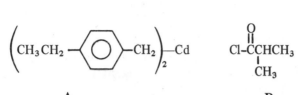

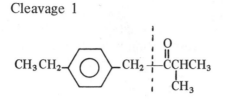

A   B

Cleavage 2

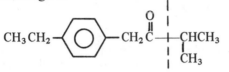

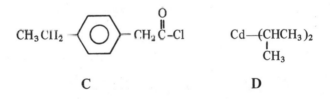

C   D

We shall use **A** and **B** as the required starting materials although **C** and **D** would work equally well. (The student may pursue the synthesis of X-b using **C** and **D** as an exercise.) **B** is an acceptable starting material so we must prepare only **A**.

(5) How can **A** be prepared?
Answer: **A** can be prepared by treating the Grignard reagent formed from **E** with CdCl$_2$ as shown in procedure X-5.

$$CH_3CH_2\!-\!\!\bigcirc\!\!-CH_2\!-\!Cl \qquad E$$

(6) How can **E** be prepared?
Answer: Since **E** is a benzyl halide, it can be prepared by treatment of the corresponding benzyl alcohol **F** with HCl as shown below.

$$CH_3CH_2-\langle\bigcirc\rangle-CH_2-OH \xrightarrow{\ HCl\ } E$$

(7) How can **F** be prepared?
Answer: **F** being a primary alcohol may be prepared by treatment of the Grignard reagent formed from **G** with CH_2O via procedure VI-4.

$$CH_3CH_2-\langle\bigcirc\rangle-Br \qquad\qquad G$$

(8) How can **G** be prepared?
Answer: **G** can be prepared by treating ethylbenzene with Br_2 and $FeBr_3$ via procedure V-3.

The synthesis of X-b now becomes

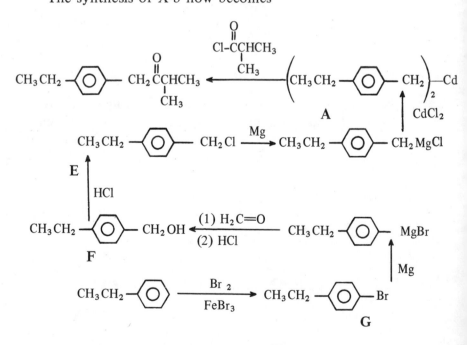

X-c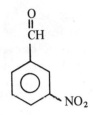

Let us ask some questions regarding X-c.

(1) What kind of molecule is X-c?
Answer: X-c is a *meta* substituted aromatic aldehyde (*m*-nitrobenzaldehyde).

(2) Do we have to add carbon atoms?
Answer: Since only the aldehyde group contains a carbon atom, we may or may not have to add carbons depending on the molecule we start with.

(3) What procedures are available for placing an aldehyde group on an aromatic ring?
Answer: Procedures X-1, X-2 and X-3.

(4) What procedures are available for placing a $-NO_2$ group on an aromatic ring?
Answer: Procedures V-6 and V-7. The use of acetyl nitrate via procedure V-7 will yield primarily an *ortho* substituted nitro group and would not be the method of choice for this preparation. Thus, we will use procedure V-6.

(5) How should we start the synthesis if we are to work it backwards?
Answer: The important item of this synthesis is the *meta* orientation of the substituents. The aldehyde group is a *meta* director, but if we attempt to nitrate benzaldehyde **A**, the aldehyde function would be oxidized up to the corresponding nitro substituted acid **B** under the conditions of the reaction.

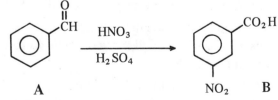

A   B

The reason for this is that benzaldehydes are very easily oxidized to the corresponding benzoic acid under the conditions of heat and acid. **B**, however, could be converted back to the corresponding aldehyde (X-c) via procedure X-2.

The use of procedure X-1 would require the nitrile **C**.

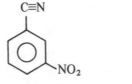

C

To attain **C**, we would have to nitrate benzonitrile **D**.

D

Let us use procedure X-1. (The student may use procedure X-2 to prepare X-c as an exercise).

Thus the synthesis of X-c becomes

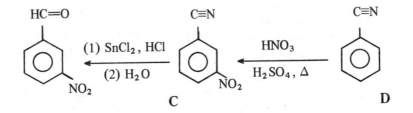

X-d   CH₃ O
      CH₃CHCHCCH₃
          CH₃

Let us ask some questions concerning X-d.

(1)   What kind of molecule is X-d?
      Answer: X-d is an aliphatic acyclic ketone (3,4-dimethyl-2-pentanone).

(2)   Do we have to add carbon atoms?
      Answer: Since X-d contains seven carbons and we are not

permitted to start with an acyclic molecule containing more than five carbons, we must add carbon atoms.

(3) What procedures are available for preparing aliphatic ketones?
Answer: Procedure X-4 and X-5. Let us use X-4. (The student may use procedure X-5 to prepare X-d as an exercise.)

(4) What is the structure of the alcohol needed so that via procedure X-4, X-d will be isolated?
Answer: The carbon which contain the ketone oxygen must contain the alcohol oxygen in the starting alcohol. Thus **A** is the required alcohol.

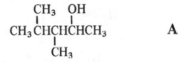

CH₃ OH positions aside:

$$\underset{\underset{\text{CH}_3}{|}}{\text{CH}_3\text{CHCHCHCH}_3} \qquad \text{A}$$

A must be synthesized since it contains seven carbons.

(5) How can **A** be prepared?
Answer: Since **A** is a secondary alcohol, it can be prepared by treating a Grignard reagent with an aldehyde via procedure VI-4-b.

(6) How can **A** be broken up so that the structures of the desired aldehyde and Grignard reagent can be deduced?
Answer: **A** can be broken up in two ways by cleaving the carbon-carbinol carbon bond as illustrated below.

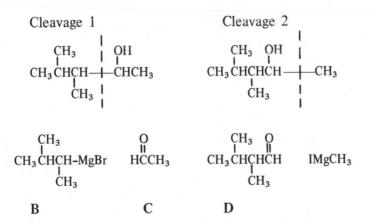

Cleavage 1   Cleavage 2

B   C   D

C and the alkyl halide from **B** do not require any further preparation. However, **D** contains six carbons and must be prepared. We shall thus use **B** and **C** since it will involve fewer steps.

Thus, the synthesis of X-d becomes

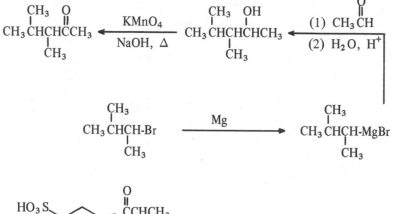

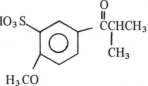

X-e

H_3CO

Let us ask some questions concerning X-e.

(1) What kind of molecule is X-e?
Answer: X-e is a disubstituted benzenesulfonic acid (2-methoxy-5-isobutyroylbenzenesulfonic acid).

(2) Do we have to add carbon atoms?
Answer: Yes, because X-e contains two substituents which contain carbon.

(3) What procedures are available for preparing an aromatic sulfonic acid?
Answer: Procedures V-8 and V-9.

(4) What procedures are available for preparing an aromatic ketone?
Answer: Procedures X-5 and X-6.

(5) What procedures are available for placing an –OCH₃ group on an aromatic ring?

Answer: Procedures VII-1 and VII-2 which require a phenol.

(6) How do we start the synthesis of X-e?
Answer: The last step of the synthesis would have two of
the substituents already on the ring. Thus, let us put the
$-SO_3H$ group using procedure V-9 on the ring last. This
would require molecule **A**.

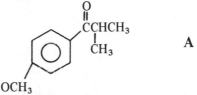

The reason that the $-SO_3H$ group is placed on the ring last
is that once on the ring the organic molecule becomes very
acidic and potentially can destroy reagents which are added
to place other groups on the ring. Note that both substituents on **A** will direct the SO_3H group to the desired
position on the ring.

(7) How can **A** be prepared?
Answer: The $-OCH_3$ group is an *ortho, para* director while
the ketone function is a *meta* director. Thus, to get the
desired *para* arrangement of the groups, the ketone function
should be added to **B** which is a permitted starting material.
Let us use procedure X-6 to incorporate the acyl group onto
the aromatic ring.

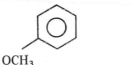

Thus the synthesis of X-e becomes:

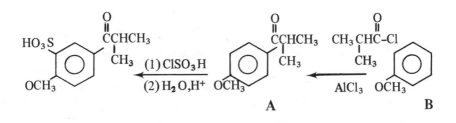

X-f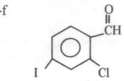

Let us ask some questions regarding X-f.

(1) What kind of molecule is X-f?
Answer: X-f is a disubstituted benzaldehyde (2-chloro-4-iodobenzaldehyde).

(2) Do we have to add carbon atoms?
Answer: Since one of the substituents contains carbon, we may or may not have to add carbons.

(3) What procedures are available for preparing aromatic aldehydes?
Answer: Procedures X-1, X-2 and X-3.

(4) What procedure is available for placing an I on an aromatic ring?
Answer: Procedure V-5.

(5) What procedure is available for placing a Cl on an aromatic ring?
Answer: Procedure V-4.

(6) What points need to be considered before we begin the synthesis of X-f?
Answer: The aldehyde function is a *meta* director and a deactivating group. The Cl and I are *ortho, para* directors and also are deactivating groups. The substitution pattern on X-f is such that the halogens are *ortho* and *para* to the aldehyde group and are *meta* to one another. The fact that the aldehyde is attached to an aromatic ring and is very susceptible to air oxidation indicates that it should be put on the ring last. The question then is how to place the aldehyde group on the ring. The use of procedure X-1 would require the conversion of a nitrile group which is a *meta* director. The use of procedure X-2 would require an acid or acid halide which are also *meta* directors. The use of procedure X-3 would require a –CH₃ group which is an

ortho para director. Thus, we shall use procedure X-3. Thus the synthesis of X-f becomes:

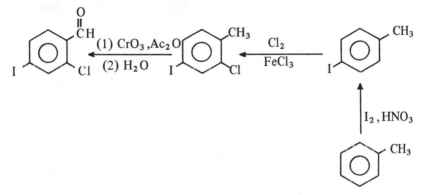

The Cl should be put on after the I is on the ring in order to insure *ortho* substitution. The I being a large atom will yield predominantly the *para* product.

X-g

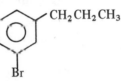

Let us ask some questions concerning X-g.

(1) What kind of molecule is X-g?
Answer: X-g is a *meta* substituted aryl halide (*m*-bromo-propylbenzene).

(2) Do we have to add carbon atoms?
Answer: Since only one of the ring substituents we may or may not have to add carbons.

(3) What procedure is available for placing a Br on an aromatic ring?
Answer: Procedure V-4.

(4) What procedure is available for placing an *n*-propyl group on an aromatic ring?
Answer: Procedure V-1, the use of $CH_3 CH_2 CH_2 Cl$ and $AlCl_3$ would yield some *n*-propyl derivature, but there would

also be some isopropyl derivative brought about by carbonium rearrangement. To avoid this rearrangement the use of procedure X-6 along with the Wolff-Kischner reaction is recommended.

(5) What points should be considered before we begin the synthesis of X-g?

Answer: Since both of the substituents are *ortho-para* directors and their arrangement on the ring is *meta*, we must have either (a) a substituent on the ring which is a *meta* director and which in turn can be converted to either a Br or a $-CH_2 CH_2 CH_3$ group or (b) a group on the ring which will permit the substitution of the Br and $-CH_2 CH_2 CH_3$ groups in a *meta* orientation and then can be removed leaving only the desired substituents. Considering alternative (a) we do not now know of a way of converting a *meta* directing group into a Br. However, using procedure X-6 along with the Wolff-Kischner reduction it is possible to place a

$-\overset{\overset{\displaystyle O}{\|}}{C}-CH_2 CH_3$ group (*meta* director) on the ring and convert it to a *n*-propyl group.

Working the synthesis backwards would require the conversion of the $-\overset{\overset{\displaystyle O}{\|}}{C}CH_2 CH_3$ to the $-CH_2 CH_2 CH_3$ in the last step. Thus, we need to prepare **A**

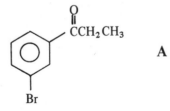

A

(6) How can **A** be prepared?

Answer: Since we require a *meta* orientation we can prepare **A** be treatment of **B** with Br_2 via procedure V-4.

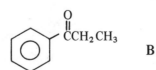

B

The synthesis of X-g now becomes:

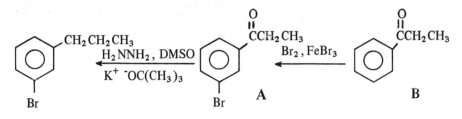

X-h

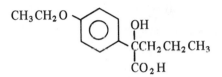

Let us ask some questions concerning X-h.

(1) What kind of molecule is X-h?
Answer: X-h is an α-hydroxy acid containing a *para* substituted aromatic ring (2 hydroxy-2(4-ethoxyphenyl)pentanoic
acid).

(2) Do we have to add carbon atoms?
Answer: Yes, because each of the substituents of the aromatic ring contain carbon and we are permitted to start
with a monosubstituted aromatic ring.

(3) What procedure is available for preparing an α-hydroxy acid.
Answer: The hydrolysis of a cyanohydrin from an aldehyde
or ketone will yield an α-hydroxy acid.

$$
\underset{\substack{R-C-R'}}{\overset{O}{\|}} \xrightarrow{\text{HCN}} \underset{\substack{CN}}{\overset{OH}{\underset{|}{R-C-R'}}} \xrightarrow[\text{(2) HCl}]{\text{(1) NaOH, H}_2\text{O}} \underset{\substack{CO_2H}}{\overset{OH}{\underset{|}{R-C-R'}}}
$$

To yield X-g we need the cyanohydrin A.

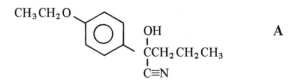

(4) How can **A** be prepared?
Answer: As demonstrated in answer (3) we can prepare **A** by treating ketone **B** with HCN

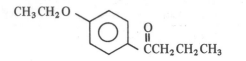

B

(5) How can **B** be prepared?
Answer: **B** can be prepared by treating phenetole with *n*-butanoyl chloride via procedure X-6.

Thus, the synthesis of X-h becomes

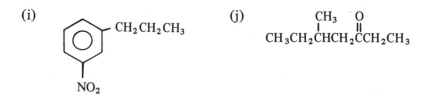

UNWORKED SYNTHESES

Using the same conditions set forth before the worked syntheses section of Chapter X, prepare the following:

(i)

$CH_2CH_2CH_3$

NO_2

(j)

$$CH_3CH_2\overset{\overset{\displaystyle CH_3}{|}}{C}HCH_2\overset{\overset{\displaystyle O}{\|}}{C}CH_2CH_3$$

(k)

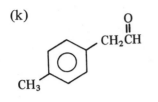

(l)   HC=O

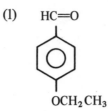

(m)

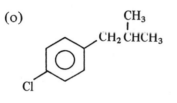

(n) $CH_3CHCH_2CH_2CH_3$
        $HC=O$

(o)

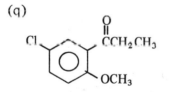

(p)   HC=O

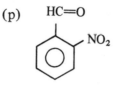

(q)

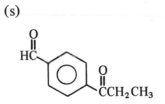

(r)

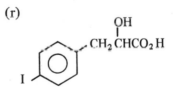

(s)

PROCEDURES FOR SYNTHESIZING AMINES

XI-1 Reduction of nitro compounds (RNO_2). This reaction can
be used to prepare aliphatic or aromatic primary amines.

(a) Catalytic method
The catalysts most frequently used are Pd, Pt and Ni.

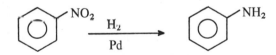

$$CH_3CHCH_3 \xrightarrow[\text{DMSO}]{NaNO_2} CH_3CHCH_3 \xrightarrow[\text{Ni}]{H_2} CH_3CHCH_3$$
$$\quad\ \ |\ \qquad\qquad\qquad\qquad |\ \qquad\qquad\qquad |$$
$$\quad\ \ Br\qquad\qquad\qquad\qquad NO_2\qquad\qquad\qquad NH_2$$

The above method cannot be used if the molecule contains
another substituent which is susceptible to catalytic
hydrogenation.

(b) Chemical method
For primary aromatic amine formation, Sn + HCl or Fe
+ HCl is most frequently used.

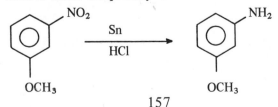

157

For primary aliphatic amines LiAlH$_4$ is the reagent most frequently used.

$$CH_3CH_2CH_2CH_2-NO_2 \xrightarrow{\text{LiAlH}_4} CH_3CH_2CH_2CH_2-NH_2$$

XI-2 Reduction of nitriles (R–C≡N)

This reaction can be used to prepare only primary aliphatic amines.

(a) Catalytic method

This reaction requires the use of H$_2$ and either Pt or Ni.

$$CH_3CH_2CH_2-Br \xrightarrow{\text{KCN}} CH_3CH_2CH_2C\equiv N \xrightarrow[\text{Pt}]{\text{H}_2} CH_3CH_2CH_2CH_2NH_2$$

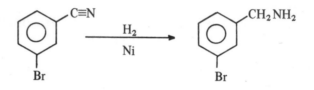

Note that the carbon of the cyano group gets converted to a –CH$_2$ – group upon reduction.

(b) Chemical method

This method requires the use of LiAlH$_4$ as the reducing material.

$$\underset{\underset{CH_3}{|}}{CH_3CH_2CHC\equiv N} \xrightarrow{\text{LiAlH}_4} \underset{\underset{CH_3}{|}}{CH_3CH_2CHCH_2NH_2}$$

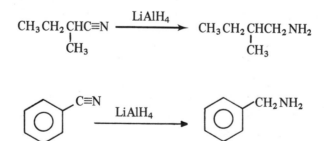

Note for methods (a) and (b) if there are other substituents on the molecule which are susceptible to reduction, they also will be reduced during the course of the reaction. $NaBH_4$ will *not* reduce $R-C \equiv N$.

XI-3 Reduction of amides

This method can be used to produce a pure primary, secondary or tertiary amine.

Note that in this reduction the $-\overset{O}{\overset{\|}{C}}-$ of the amide becomes converted to a $-CH_2-$. Thus, this reduction will have utility in forming amines only in which there is a $-CH_2-$ group attached to the N.

Formation of a primary amine: requires a primary amide.

Formation of a secondary amine: requires a secondary amide.

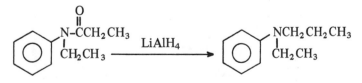

Formation of a tertiary amine: requires a tertiary amide.

XI-4 Hofmann reaction

This reaction may be used to form primary, secondary and tertiary aliphatic and aromatic amines. The reaction requires a primary or secondary alkyl halide. If an aryl halide is used, the halogen must be *ortho* or *para* to strong electron withdrawing groups.

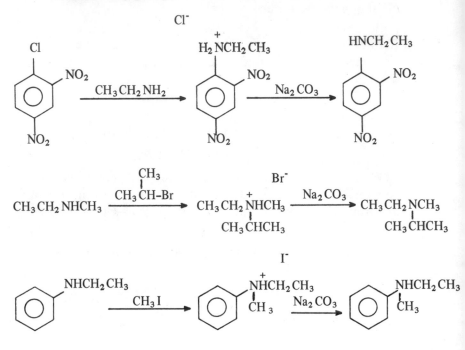

WORKED SYNTHESES

Using any monosubstituted cyclic organic molecule, any mono-substituted acyclic molecule containing no more than five carbons and any inorganic reagents prepare the following.

XI-a $\quad$ CH_3CH_2
$\quad CH_3CH_2CHNCH_2CH_3$
$\qquad CH_2CH_3$

Let us ask some questions concerning XI-a.

(1) What kind of molecule is XI-a?
Answer: XI-a is a tertiary aliphatic amine (N,N-diethyl-3-aminopentane).

(2) Do we have to add carbon atoms?
Answer: Yes, because XI-a contains carbon atoms while we
are permitted to start with acyclic molecules containing no
more than five carbons.

(3) What procedures are available for preparing tertiary aliphatic
amines?
Answer: Procedures XI-3 requiring an amide and XI-4 which
requires an alkyl halide and a secondary amine. We shall
use procedure XI-4.

(4) How can XI-a be broken up so that the structures of the
alkyl halide and secondary amine can be deduced?
Answer: Since XI-a is a tertiary amine, it can be broken
down into its required starting materials in three ways which
require the cleavage of a C–N bond as illustrated below.

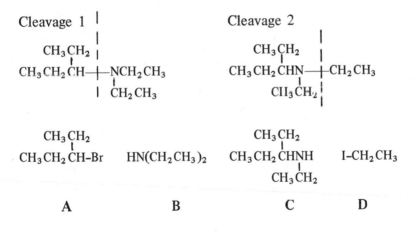

Cleavage 1

Cleavage 2

A   B   C   D

Cleavage 3

Note that cleavages 2 and 3 require the same starting
materials; thus for this particular problem there are only
two ways of forming XI-a using procedure XI-4. To decide

whether to use cleavage 1 or 2, let us look at the starting materials. **A** and **B** can be used without further preparation because they are each monosubstituted compounds which do not contain more than five carbons. **C**, on the other hand, must be prepared because it is a monosubstituted molecule containing seven carbons, two more than we are permitted to use. Thus, the shortest method would involve using **A** and **B**.

The synthesis of XI-a now becomes

$$CH_3CH_2CH_2CHNCH_2CH_3 \text{ with } CH_3CH_2 \text{ and } CH_2CH_3 \xleftarrow{Na_2CO_3} CH_3CH_2CHNCH_2CH_3 \text{ with } CH_3CH_2 \text{ and } Br^- \text{ and } CH_2CH_3$$

$$\uparrow$$

$$CH_3CH_2NHCH_2CH_3 \qquad \textbf{B}$$

$$CH_3CH_2CHCH_2CH_3$$
$$\underset{Br}{|}$$

$$\textbf{A}$$

XI-b

OCH$_3$

NHCH$_2$CHCH$_3$
$\underset{CH_3}{|}$

Let us ask some questions concerning XI-b.

(1) What kind of molecule is XI-b?
Answer: XI-b is a secondary aromatic amine (N-isobutyl-2-methoxyaniline).

(2) Do we have to add carbon atoms?
Answer: Since the aromatic ring is disubstituted and each substituent contains carbon atoms, we must add carbon atoms.

(3) What procedures are available for preparing secondary aromatic amines?

Answer: Procedures XI-3, which requires an amide and
that the desired amine contains the grouping N–CH$_2$–, and
procedure XI-4, which is the Hofmann reaction. Let us use
procedure XI-3. The student may prepare XI-b via procedure
XI-4 as an exercise.

(4)   What is the structure of the amide required so that upon reduction XI-b will result?
Answer: Since the –CH$_2$– group adjacent to the N in the

final product must be a $-\overset{\overset{\displaystyle O}{\|}}{C}-$ in the amide, the structure of
the required amide is A.

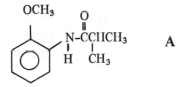

(5)   How can A be prepared?
Answer: Since A is a secondary amide, it can be prepared
by treating an amine with an acid chloride via procedure
IX-7.

(6)   How can A be cleaved so that the structure of the required
amine and acid chloride can be determined?
Answer: To determine the structures of the required start-

ing materials the N–$\overset{\overset{\displaystyle O}{\|}}{C}$ must be cleaved as illustrated below.

B   C

C is a permitted starting material since it is a monosubstituted organic molecule containing less than five carbons.
B on the other hand must be synthesized.

(7) How can **B** be prepared?
Answer: **B** is a primary aromatic amine and thus can be
prepared by procedure XI-1 which would require **D**.

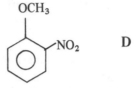

OCH₃

NO₂   **D**

(8) How can **D** be prepared?
Answer: **D** is a molecule which contains an *ortho* nitro
group. This group can most easily be prepared by treating
anisole with acetyl nitrate via procedure V-7. Thus the
synthesis of XI-b becomes

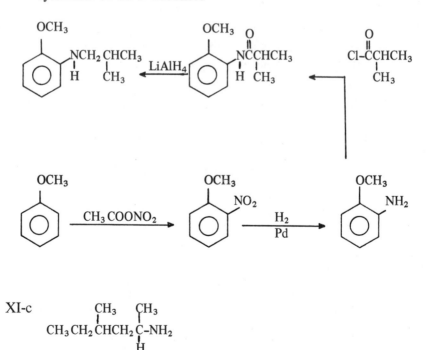

XI-c   CH₃   CH₃
        CH₃CH₂CHCH₂C-NH₂
                      |
                      H

Let us ask some questions concerning XI-c.

(1) What kind of molecule is XI-c?

Answer: XI-c is a primary aliphatic amine (2-amino-4-methylhexane).

(2) Do we have to add carbon atoms?
Answer: Since XI-c contains seven carbons and we are not permitted to start with any monosubstituted molecule containing more than five carbons, we must add carbon atoms.

(3) What procedures are available for preparing a primary aliphatic amine?
Answer: Procedures XI-1, XI-2, XI-3 and XI-4. However, procedures XI-2 and XI-3 will produce an amine with the structure $-CH_2NH_2$. Since XI-c does not contain this functional grouping, we must choose between procedures XI-1 and XI-4. Let us select XI-1. (As an exercise the student may use procedure XI-4 to prepare XI-c). Thus we need **A** which upon reduction will yield XI-c.

$$CH_3CH_2\overset{\overset{\displaystyle CH_3}{|}}{C}H CH_2\overset{\overset{\displaystyle CH_3}{|}}{C}HNO_2 \qquad \textbf{A}$$

(4) How can **A** be prepared?
Answer: An aliphatic nitro compound may be synthesized by treating an alkyl halide with $NaNO_2$ and DMSO as shown in procedure XI-1. Thus molecule **B** is required as a starting material for **A**.

$$CH_3CH_2\overset{\overset{\displaystyle CH_3}{|}}{C}H CH_2\overset{\overset{\displaystyle CH_3}{|}}{C}HBr \qquad \textbf{B}$$

(5) How can **B** be prepared?
Answer: A method of converting primary or secondary alcohols directly to the corresponding alkyl bromide involves the treatment of the alcohol with PBr_3 as shown below. This reaction is useful because carbon skeleton rearrangement does not occur.

$$R\text{-}OH + PBr_3 \longrightarrow R\text{-}Br$$

Thus, if we desire to prepare **B** which is a secondary bromide we merely have to react the corresponding alcohol **C** with PBr_3.

$$\underset{\substack{CH_3\\|}}{} \underset{\substack{CH_3\\|}}{}$$

$$CH_3CH_2\overset{CH_3}{\underset{|}{C}}HCH_2\overset{CH_3}{\underset{|}{C}}H-OH \qquad \textbf{C}$$

It should be noted that **B** could also be prepared by adding HBr to the correct olefin.

(6) How can **C** be prepared?

Answer: **C** being a secondary alcohol can be prepared by reacting a Grignard reagent to an aldehyde according to procedure VI-4b. Since **C** is a secondary alcohol, it can be broken up in two ways, namely by cleaving each of the carbinol carbon-carbon bonds as illustrated below.

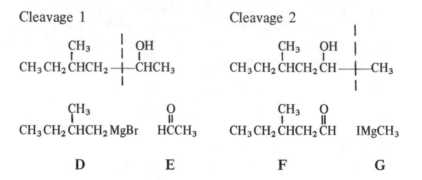

Cleavage 1

$$CH_3CH_2\overset{CH_3}{\underset{|}{C}}HCH_2\underset{|}{\overset{\ \ |\ \ OH}{|\ \ }}CHCH_3$$

$$CH_3CH_2\overset{CH_3}{\underset{|}{C}}HCH_2MgBr \qquad \overset{O}{\underset{||}{H}}CCH_3$$

D   **E**

Cleavage 2

$$CH_3CH_2\overset{CH_3}{\underset{|}{C}}HCH_2\overset{OH}{\underset{}{C}}H\underset{|}{\overset{|}{\ \ }}CH_3$$

$$CH_3CH_2\overset{CH_3}{\underset{|}{C}}HCH_2\overset{O}{\underset{||}{C}}H \qquad IMgCH_3$$

F   **G**

Since cleavage 1 yields **D** and **E** each of which do not contain more than five carbons, while cleavage 2 yields **F** which contains six carbons and would have to be prepared, thus leading to more steps, let us select **D** and **E** as the shortest route to the preparation of **C**.

Thus the synthesis of XI-c becomes

$$CH_3CH_2\overset{CH_3}{\underset{|}{C}}HCH_2\overset{CH_3}{\underset{|}{C}}H-NH_2 \xleftarrow[Pt]{H_2} CH_3CH_2\overset{CH_3}{\underset{|}{C}}HCH_2\overset{CH_3}{\underset{|}{C}}H-NO_2 \xleftarrow[DMSO]{NaNO_2}$$

$$\overset{(1)\ CH_3\overset{O}{\overset{||}{C}}CH_3}{\underset{(2)\ H_2O,\ H^+}{}} CH_3CH_2\overset{CH_3}{\underset{|}{C}}HCH_2\overset{CH_3}{\underset{|}{C}}H-OH \xrightarrow{PBr_3} CH_3CH_2\overset{CH_3}{\underset{|}{C}}HCH_2\overset{CH_3}{\underset{|}{C}}H-Br$$

$$CH_3CH_2\overset{CH_3}{\underset{|}{C}}HCH_2MgBr \xleftarrow{Mg} CH_3CH_2\overset{CH_3}{\underset{|}{C}}HCH_2-Br$$

XI-d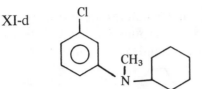

Let us ask some questions concerning XI-d.

(1) What kind of molecule is XI-d?
Answer: XI-d is tertiary aromatic amine (N -methyl-N -cyclo-hexyl-*m*-chloroaniline).

(2) Do we have to add carbons?
Answer: Yes, because the amino function is an *ortho para* director and thus the *meta* location of the Cl dictates that the amine function will have to be placed on the ring after the Cl. Thus the cyclohexyl group will have to be added to the aromatic molecule.

(3) What procedures are available for preparing tertiary aromatic amines?
Answer: Procedures XI-3 and XI-4. However, since XI-d does not possess the functional grouping –CH₂ N–, procedure XI-3 cannot be utilized. Thus, procedure XI-4 is the method of choice.

(4) How can XI-d be broken up so that the structures of the re-quired starting materials can be deduced?
Answer: Being a tertiary amine, there are three C–N bonds which may be cleaved as illustrated below.

Cleavage 1   Cleavage 2   Cleavage 3

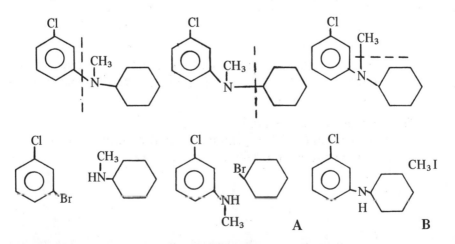

Cleavage 1 is poor because aromatic halides are unreactive unless strong electron withdrawing groups are located *ortho* or *para* to the halide. Cleavages 2 and 3 are both possible. However, CH_3 is smaller than ⟨ ⟩ , thus it probably would react faster. Thus, let us use the starting materials **A** and **B**.

(5) How can **A** be prepared?
Answer: **A** is a secondary aromatic amine and thus can be prepared via procedure XI-4.

(6) How can **A** be broken up so that the starting materials necessary for synthesizing **A** via procedure XI-4 can be deduced?
Answer: Being a secondary amine **A** can be broken up by cleaving either of the two C–N bonds as illustrated below.

Cleavage 4   Cleavage 5

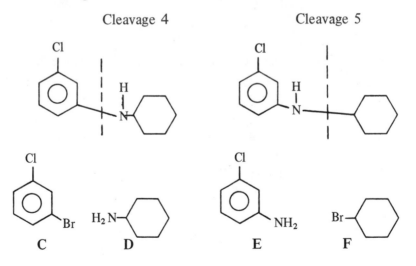

For reasons stated in answer (4), **C** is not a reactive enough halide. Thus, **E** and **F** are the required starting materials for the synthesis of **A**.

(7) How can **E** be prepared?
Answer: Since **E** is a primary aromatic amine, it can be prepared via procedure XI-1 by reducing **G**.

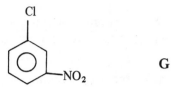

G

(8) How can **G** be prepared?

Answer: The *meta* arrangement of the Cl and NO_2 groups indicate that we must chlorinate nitrobenzene via procedure V-4.

Thus, the synthesis of XI-d becomes

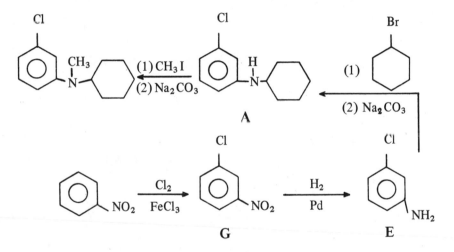

XI-e

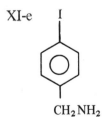

Let us ask some questions concerning XI-e.

(1) What kind of molecule is XI-e?

Answer: XI-e is an aliphatic primary amine (*p*-iodobenzylamine).

(2) Do we have to add carbon atoms?
Answer: Since only one of the substituents on the ring contains carbon, we may or may not have to add carbon atoms.

(3) What procedures are available for preparing primary aliphatic amines?
Answer: Procedures XI-1, XI-2, XI-3 and XI-4. It is noteworthy that procedures XI-2 and XI-3 yield the $-CH_2 NH_2$ grouping which XI-e possesses. However, the I can be replaced by H if $LiAlH_4$ is used as the reducing agent. Thus, the best ways of preparing the amine would involve the use of procedures XI-1 and XI-4. Let us select XI-1. This would require A.

A

(4) How can A be prepared?
Answer: A can be prepared by treating B with I_2 via procedure V-5.

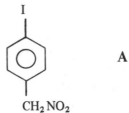

B

(5) How can B be prepared?
Answer: B can be prepared by treating C with $NaNO_2$ and DMSO as illustrated in procedure XI-1.

C

(6) How can **C** be prepared?
Answer: **C** can be prepared by treating **D** with PBr₃ as illustrated in problem XI-c.

D

CH₂OH

(7) How can **D** be prepared?
Answer: **D** being a primary alcohol can be prepared by reacting the Grignard reagent **E** with formaldehyde.

E

MgI

Thus the synthesis of XI-e becomes

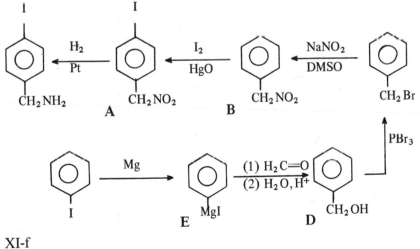

XI-f

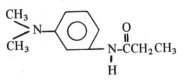

Let us ask some questions concerning XI-f.

(1)   What kind of molecule is XI-f?
Answer: XI-f is secondary aromatic amide also containing a tertiary aromatic amine function [N‑(3‑N',N'-dimethylamino-phenyl)propanamide].

(2)   Do we have to add carbon atoms?
Answer: Since the ring is disubstituted and each substituent of the ring contains carbon, we do have to add carbon atoms.

(3)   What procedure is available for preparing a secondary aromatic amide?
Answer: Procedure IX-7 which requires a primary aromatic amine and an acid chloride.

(4)   What procedures are available for preparing a tertiary aromatic amine?
Answer: Procedures XI-3 and XI-4.   Since procedure XI-3 yields a product with the functional grouping $-CH_2NH_2$, this eliminates this procedure for use in preparing XI-f. Thus, we have to use procedure XI-4.

(5)   How do we begin the synthesis of XI-f?
Answer: In the last step of the reaction we can either acylate **A** using procedure IX-7 or place the CH_3 groups on **B** using procedure XI-4.

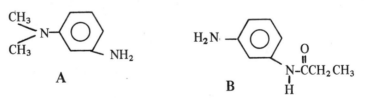

Let us use the former procedure.   Thus, we must prepare **A**.

(6)   How can **A** be prepared?
Answer: **A** being a primary amine can be prepared from **C** via procedure XI-1.

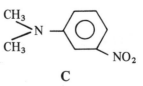

(7) How can **C** be prepared?
Answer: **C** being a tertiary aromatic amine can be prepared by treating **D** with $CH_3 I$ (2 moles) via procedure XI-4.

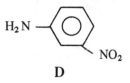

D

(8) How can **D** be prepared?
Answer: **D** can be prepared by treating **E** with $(NH_4)_2 S$. This reaction is good for the reduction of one of two NO_2 groups which are arranged in a *meta* orientation on an aromatic ring.

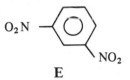

E

(9) How can **E** be made?
Answer: **E** can be made by nitrating nitrobenzene via procedure V-6.

Thus the synthesis of XI-f becomes

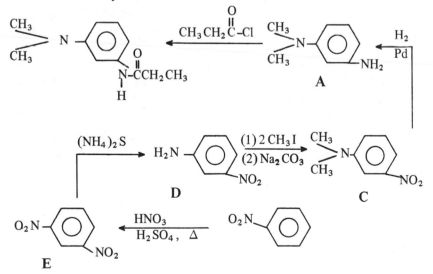

XI-g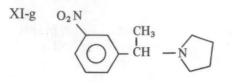

Let us ask some questions concerning XI-g

(1) What kind of molecule is XI-g?
Answer: XI-g is an aliphatic cyclic tertiary amine (α-pyrrolidino-*m*-nitroethylbenzene).

(2) Do we have to add carbon atoms?
Answer: Since it would be unfeasible to place the nitro group on the aromatic ring in the *meta* position if we started with the rest of the molecule intact, we will have to add carbon atoms.

(3) What procedures are available for preparing a tertiary aromatic amine?
Answer: Procedures XI-3 and XI-4. It should be noted that procedure XI-3 will produce tertiary amines with the general grouping $-CH_2-N-$. Since XI-g does not contain that grouping, we will have to use procedure XI-4 to produce XI-g.

(4) How can XI-g be broken up so that the required starting materials needed to produce XI-g via procedure XI-4 can be discerned?
Answer: Being a tertiary amine, there are three C–N bonds which can be cleaved to show the required starting materials. However, in cases where the N is part of a ring system, it is best not to cleave a ring C–N bond. This leaves only one C–N which when cleaved yields the starting materials **A** and **B**.

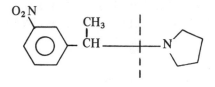

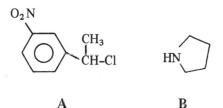

A   B

Note **B** is a permitted starting material. We must, however, prepare **A**.

(5) How can **A** be prepared?
Answer: **A** is a secondary benzyl chloride. It, thus, can be prepared by treating the alcohol **C** with HCl.

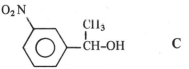

(6) How can **C** be prepared?
Answer: **C** being a secondary alcohol can be prepared by reducing **D** via procedure VI-3b. $NaBH_4$ will not reduce the $-NO_2$ group.

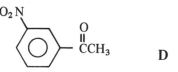

Note that a Grignard reaction will not work because of the presence of the $-NO_2$ group.

(7) How can **D** be prepared?
Answer: Since **D** is an aromatic ketone, the best procedure which is available for placing an acyl group on an aromatic

ring involves procedure X-6. However, the $-NO_2$ group on the ring prohibits the Friedel-Crafts reaction from occurring. Thus, we will have to nitrate the ring already containing the $-\overset{\overset{\displaystyle O}{\|}}{C}-CH_3$ group. This is feasible and will yield **D**. Thus we must begin with acetophenone **E**. The synthesis of XI-g now becomes

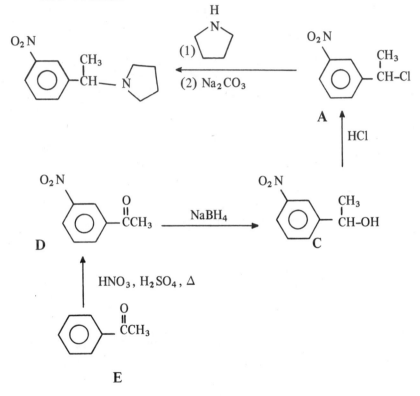

UNWORKED SYNTHESES

Synthesize the following using the same instructions specified for the worked syntheses problems.

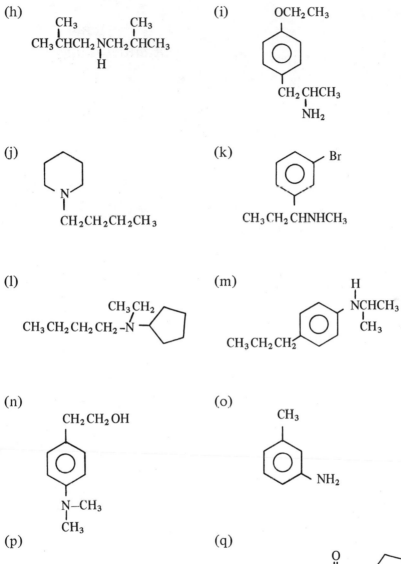

(h)

CH₃CHCH₂NCH₂CHCH₃
with CH₃ groups and H on nitrogen

(i)

OCH₂CH₃

CH₂CHCH₃
NH₂

(j)

N
CH₂CH₂CH₂CH₃

(k)

Br

CH₃CH₂CHNHCH₃

(l)

CH₃CH₂
CH₃CH₂CH₂CH₂–N
cyclopentyl

(m)

H
NCHCH₃
CH₃

CH₃CH₂CH₂

(n)

CH₂CH₂OH

N–CH₃
CH₃

(o)

CH₃

NH₂

(p)

CH₂NH₂

–NH₂

(q)

O
N–C cyclopentyl
H

CH₃CH₂NCH₂CH₃

(r)

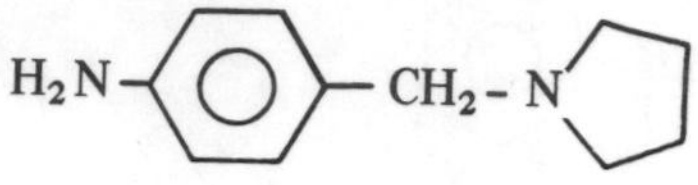

$$H_2N-\langle\bigcirc\rangle-CH_2-N\langle\ \rangle$$

(s)

$$CH_3CH_2CH_2\overset{\overset{\displaystyle CH_3}{|}}{\underset{\underset{\displaystyle CH_3}{|}}{C}}-NH_2$$

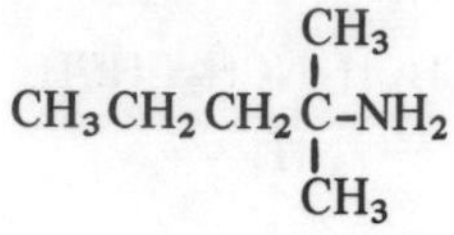

(Hint: use the Ritter reaction.)

CHAPTER XII

AROMATIC DIAZONIUM SALTS

PROCEDURES FOR SYNTHESIZING AROMATIC DIAZONIUM SALTS

XII-1   Preparation of aromatic diazonium salts
Although the literature contains several methods for preparing aromatic diazonium salts the most commonly used
technique involves the treatment of an aromatic primary
amine with sodium nitrite ($NaNO_2$) and a strong acid
(HCl, HBr, H_2SO_4 are the most frequently used).

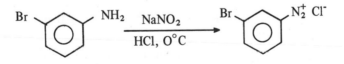

$$Br \text{—} \underset{}{\bigcirc} \text{—} NH_2 \xrightarrow[\text{HCl, 0°C}]{NaNO_2} Br \text{—} \underset{}{\bigcirc} \text{—} N_2^+ \, Cl^-$$

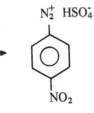

$$\underset{NO_2}{\overset{NH_2}{\bigcirc}} \xrightarrow[\text{H}_2\text{SO}_4, \, 0°\text{C}]{NaNO_2} \underset{NO_2}{\overset{N_2^+ \, HSO_4^-}{\bigcirc}}$$

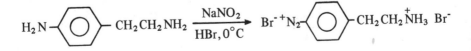

$$H_2N \text{—} \underset{}{\bigcirc} \text{—} CH_2CH_2NH_2 \xrightarrow[\text{HBr, 0°C}]{NaNO_2} Br^- \, ^+N_2 \text{—} \underset{}{\bigcirc} \text{—} CH_2CH_2 \overset{+}{N}H_3 \, Br^-$$

179

There are several points which must be understood if one is going to prepare an aromatic diazonium salt.

(1)   In nearly all the cases the salt is prepared and at ice bath temperatures.

(2)   The groups, –R, –NO$_2$, –C̈–H, –X, –C̈–R, –CO$_2$H,

–CN and –SO$_3$H do not interfere with the diazotization of an aromatic amine. An aliphatic amine function may even be present and not interefere provided that the pH of the solution is less than 3.

CONVERSION OF THE DIAZONIUM GROUP INTO OTHER FUNCTIONAL GROUPS VIA REPLACEMENT REACTIONS

XII-2   Replacement by the fluoride function
This reaction requires NaBF$_4$ (sodium fluoborate).

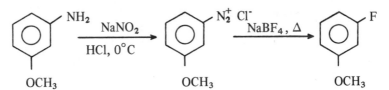

XII-3   Replacement by the chloride or bromide function (Cl or Br).
The Sandmeyer reaction requires the aromatic diazonium salt and cuprous halide (CuX).

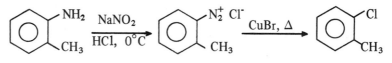

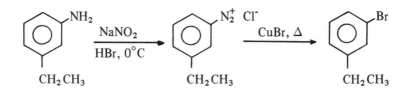

XII-4 Replacement by the iodide function (I).

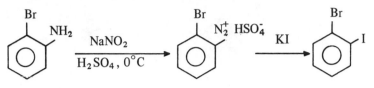

XII-5 Replacement by the phenol function (OH).

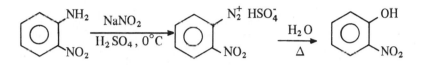

For best results the diazonium salt should be prepared by using H_2SO_4. This prevents the incorporation of Cl or Br if HCl or HBr is used in the diazotization step.

XII-6 Replacement by the alkoxy group (OR).

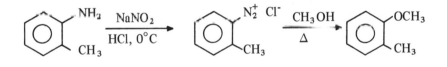

XII-7 Replacement by the cyano function (CN).

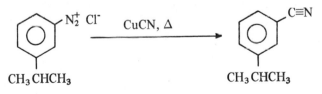

XII-8 Replacement by the nitro group (NO_2).

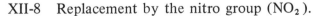

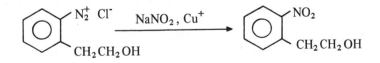

XII-9   Replacement by the azide function (N_3).

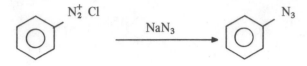

XII-10. Replacement by hydrogen (H).
This reaction requires hypophosphorous acid (H_3PO_2).

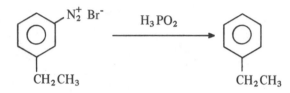

WORKED SYNTHESES

Using any monosubstituted acyclic organic molecule containing no more than four carbons, any monosubstituted cyclic organic molecule, and any inorganic reagents prepare the following.

XII-a

Let us ask some questions concerning XII-a.

(1)   What kind of molecule is XII-a?
Answer: XII-a is an *ortho* disubstituted aromatic ring (1,2-dinitrobenzene).

(2)   What methods are available for placing $-NO_2$ groups on aromatic rings?
Answer: Procedures V-6, V-7 and XII-8. Procedures V-6 and V-7 insert a $-NO_2$ group directly onto an aromatic ring. However, if one nitrates nitrobenzene the major product is the *meta* isomer. Thus, the only other available method would be the use of procedure XII-8.

(3) What is the structure of the compound which will yield XII-a via procedure XII-8?
Answer: The diazonium salt **A**.

N_2^+ Cl⁻

A

NO₂

(4) How can **A** be prepared?
Answer: **A** can be prepared by diazotizing the amine **B** via procedure XII-1.

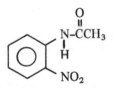

NH₂

B

NO₂

(5) How can **B** be prepared?
Answer: By hydrolyzing Cwith NaOH.

O
‖
N–CCH₃
|
H

C

NO₂

(6) How can **C** be prepared?
Answer: The –NHCOCH₃ group is an *ortho, para* directing group. Thus, treating acetanilide with acetyl nitrate via procedure V-7 one can obtain **C**. Thus, the synthesis of XII-a becomes

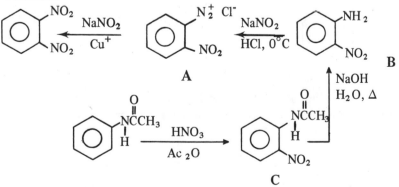

XII-b

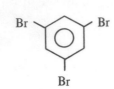

Let us ask some questions concerning XII-b.

(1)   What kind of molecule is XII-b?
Answer: XII-b is a tribromo benzene (1,3,5-tribromobenzene).

(2)   What procedures are available for placing a Br on an aro-matic ring?
Answer: Procedure V-4 which involves the electrophilic sub-stitution of the aromatic ring using Br_2 and $FeBr_3$ and pro-cedure XII-3 which requires a diazonium salt and CuBr.

(3)   How do we begin the synthesis of XII-b?
Answer: The Br atom is an *ortho para* directing group. However, the Br atoms on XII-b are all *meta* to one another. It is not readily apparent, but if one of the positions occu-pied by a hydrogen in XII-b is a strong *ortho para* director instead of hydrogen, the Br atoms could be placed on the ring in the correct positions using procedure XII-3. It would then be necessary to remove the *ortho para* directing group. The $-NH_2$ group is a strong *ortho para* director and can easily be removed via procedure XII-10. Thus in the last step of the synthesis compound **A** would be required.

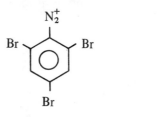

A

(4)   How can **A** be prepared?
Answer: By diazotization of **B** via procedure XII-1.

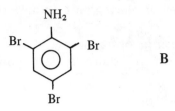

B

(5) How can **B** be prepared?
Answer: By reaction of aniline with Br_2.
Thus the synthesis of XII-b becomes

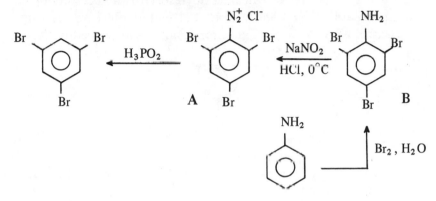

XII-c

Let us ask some questions concerning XII-c.

(1) What kind of molecule is XII-c?
Answer: XII-c is a *meta* substituted aromatic dihalide (*m*-chloroiodobenzene).

(2) What procedures are available for placing a Cl on an aromatic ring?
Answer: Procedure V-4, which requires Cl_2 and $FeCl_3$ and procedure XII-3, which utilizes a diazonium salt and CuCl.

(3) What procedures are available for placing an I on an aromatic ring?

Answer: Procedure V-5, which involves the use of I_2 and HNO_3 and procedure XII-4, which requires a diazonium salt and KI.

(4) How do we begin the synthesis of XII-c?

Answer: The *meta* arrangement on the ring of the two halides, which are both strong *ortho para* directors suggests that the uses of procedures V-4 and V-5 for placing the Cl or I on the aromatic ring in the presence of the other halide will fail. Thus we will have to resort to the diazonium salt procedures XII-3 and XII-4. Let us put the I on the ring last. We will then need compound **A** and procedure XII-4.

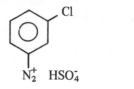

A

(5) How can **A** be prepared?

Answer: By treatment of **B** with $NaNO_2$ and H_2SO_4 via procedure XII-1.

B

(6) How can **B** be prepared?

Answer: By reducing **C** via procedure XI-1.

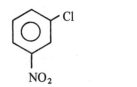

C

(7) How can **C** be prepared?

Answer: By chlorinating nitrobenzene via procedure V-4. Thus, the synthesis of XII-c becomes

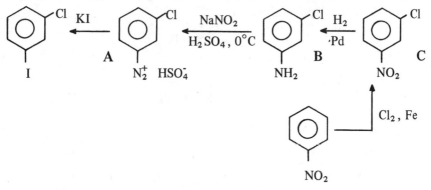

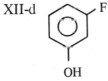

XII-d

Let us ask some questions concerning XII-d.

(1) What kind of molecule is XII-d?
Answer: XII-d is a *meta* substituted phenol (3-fluorophenol).

(2) What procedure is available for placing a F on an aromatic ring?
Answer: Procedure XII-2 which requires a diazonium salt and $NaBF_4$.

(3) What procedures are available for placing an –OH group on an aromatic ring?
Answer: Procedure XII-5 which involves a diazonium salt and a procedure involving the cleavage of an aromatic ether as illustrated below.

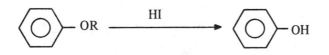

(4) How do we begin the synthesis of XII-d?
Answer: Since both substituents can be placed on the aromatic ring via diazonium salts, let us pursue that method.

We shall place the –OH group on the ring last. Thus, we shall need compound **A**.

N_2^+ HSO_4^-

A

(5) How can **A** be prepared?
Answer: By diazotization of **B** via procedure XII-1.

NH$_2$

B

(6) How can **B** be prepared?
Answer: By reducing **C** via procedure XI-1.

NO$_2$

C

(7) How can **C** be prepared?
Answer: By reacting **D** with NaBF$_4$ via procedure XII-2.

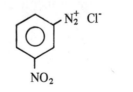

N_2^+ Cl$^-$

NO$_2$

D

(8) How can **D** be prepared?
Answer: By diazotization of **E** via procedure XII-1.

NH$_2$

NO$_2$

E

(9) How can **E** be prepared?

Answer: By reducing **F** with $(NH_4)_2 S$ as shown in problem XI-f.

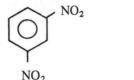

F

(10) How can **F** be prepared?

Answer: By nitration of nitrobenzene.

Thus the synthesis of XII-d becomes

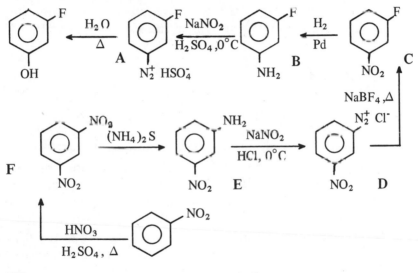

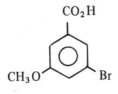

XII-e

Let us ask some questions concerning XII-e.

(1) What kind of molecule is XII-e?

Answer: XII-e is a disubstituted benzoic acid (3-bromo-5-methoxybenzoic acid).

(2) What procedures are available for placing a $-CO_2H$ group on an aromatic ring?
Answer: Procedure VII-1 involving the carbonation of a Grignard reagent; procedure VII-2 which is the hydrolysis of a nitrile; and procedure VIII-3, which is the oxidation of an alkyl benzene.

(3) What procedures are available for placing a Br on an aromatic ring?
Answer: Procedure V-4, which requires Br_2 and $FeBr_3$ and procedure XII-3 which involves a diazonium salt and CuBr.

(4) What procedure is available for placing an $-OCH_3$ group on an aromatic ring?
Answer: Procedure XII-6 which requires a diazonium salt and procedure VII-1, the Williamson synthesis in which a

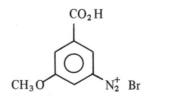 is required.

(5) How do we begin the synthesis of XII-e?
Answer: The *meta* orientation of all the groups on the ring is important. The only functional group in XII-e which is a *meta* director is the $-CO_2H$ group. Since we are working the synthesis by starting with the desired product and proceeding step-wise back to a permitted starting material, we normally place on the ring in the last step the functional group which can be incorporated onto the ring in the fewest ways. In the case of XII-e the $-OCH_3$ and the Br can each be placed on the ring using two procedures. Thus, we can place the Br or the $-OCH_3$ group on the ring in the last step. If we choose to place the Br on the ring in the last step this will require compound **A** and procedure XII-3.

A

(6) How can **A** be prepared?
Answer: By diazotization of **B** via procedure XII-1.

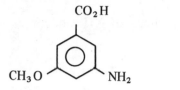

B

(7) How can **B** be prepared?
Answer: By reduction of **C** via procedure XI-1.

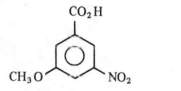

C

(8) How can **C** be prepared?
Answer: By reaction of **D** with CH_3OH via procedure XII-7.

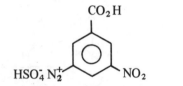

D

(9) How can **D** be prepared?
Answer: By diazotization of **E** via procedure XII-1.

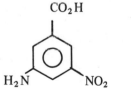

E

(10) How can **E** be prepared?
Answer: By reduction of **F** as illustrated in problem XI-f.

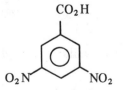

(11) How can **F** be prepared?
Answer: By nitration of benzoic acid using fuming HNO_3 and H_2SO_4.
Thus, the synthesis of XII-e becomes

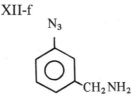

XII-f

Let us ask some questions concerning XII-f.

(1) What kind of molecule is XII-f?
Answer: XII-f is a *meta* substituted benzyl amine (3-azido-benzylamine).

(2) What procedures are available for placing a $-CH_2NH_2$ group on an aromatic ring?

Answer: The fastest method involves the reduction of an aromatic nitrile via procedure XI-2. Two other procedures XI-1 requiring an aliphatic nitro molecule and XI-3 involving the reduction of an amide would also work.

(3)   What procedure is available for placing an azide function on an aromatic ring?
Answer: Procedure XII-9 which requires a diazonium salt.

(4)   How do we begin the synthesis of XII-f?
Answer: The benzyl amine function is an *ortho para* director while the azide group could be considered a *meta* director based on its unsaturated structure. The *meta* substitution pattern of XII-f would suggest that we could nitrate azido-benzene, form the diazonium salt and run a Sandmeyer reaction to yield the desired nitrile. Reduction would yield the desired benzyl amine. However, the thermal stability of the azide linkage as well as its reactivity to many reagents strongly mitigates against this path. Because of the instability of $-N_3$, it would be best that this group be put on in the last step of the synthetic sequence. Thus, this implies that compound A and procedure XII-9 would be the last step of the synthesis.

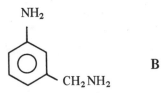

(5)   How can **A** be prepared?
Answer: By diazotization of **B** via procedure XII-1. Note that it is possible to diazotize an aromatic amine in the presence of an aliphatic amine provided the pH of the solution is kept below 3.

(6) How can **B** be prepared?
Answer: By reduction of **C** via procedure XI-1.

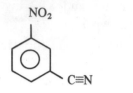

C

(7) How can **C** be prepared?
Answer: By nitration of benzonitrile via procedure V-6.

Thus the synthesis of XII-f becomes

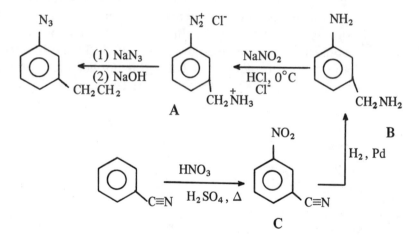

XII-g

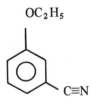

Let us ask some questions concerning XII-g.

(1) What kind of molecule is XII-g?
Answer: XII-g is a *meta* substituted benzonitrile (3-ethoxy-benzonitrile).

(2) What procedures are available for placing ether groups on an aromatic nucleus?
Answer: Procedure XII-6 which requires the use of a diazonium salt and procedure VII-1, the Williamson synthesis which requires a sodium phenoxide and an alkyl halide.

(3) What procedures are available for placing a –CN group on an aromatic ring?
Answer: Procedure XII-7 which requires a diazonium salt, and the Rosemund-von Braun reaction in which an aromatic halide is treated with cuprous cyanide as shown below.

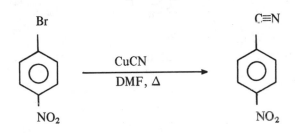

(4) How should we begin the synthesis of XII-g?
Answer: The *meta* substitution pattern of XII-g looms as the major difficulty in this synthesis. The ethoxy group is an *ortho para* director while the cyano function is a *meta* director. It appears that an easy approach to take would be to nitrate benzonitrile, reduce the –NO$_2$ group to the amine diazotize, and treat the diazonium salt with ethanol. However, it is not feasible to reduce the –NO$_2$ group of *meta*-nitrobenzonitrile without also reducing the –CN group. Thus another approach has to be undertaken. A better approach involves treatment of **A** with CuCN via the Rosemund-von Braun reaction.

A

(5) How can **A** be prepared?
   Answer: By treatment of **B** with ethanol via procedure XII-6.

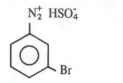

B

(6) How can **B** be prepared?
   Answer: By diazotization of **C** via procedure XII-1.

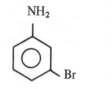

C

(7) How can **C** be prepared?
   Answer: By reduction of **D** via procedure XI-1.

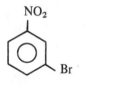

D

(8) How can **D** be prepared?
   Answer: By bromination of nitrobenzene via procedure V-9.
   Thus, the synthesis of XII-g becomes

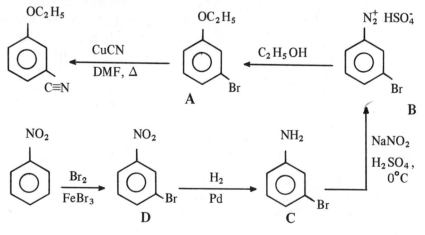

UNWORKED SYNTHESES

Following the instructions preceding the worked synthesis section, prepare the following:

(h)
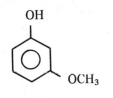

(i)
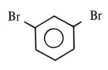

(j)
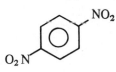

(k)

(l)
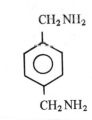

(m)

(n)
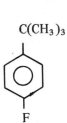

(o)

(p)

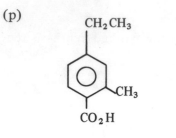

(q)

(r)

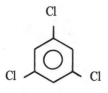

(s)

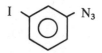

(t)

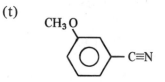

PROCEDURES FOR SYNTHESIZING PHENOLS

XIII-1 Oxidation of arylthallium compounds

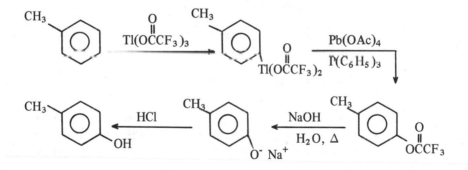

Note: The thallium functional group is believed to proceed via an electrophilic substitution mechanism. However, the position of attack on the aromatic ring by the
Tl group is not as predictable as it was for other electro-
philes. For –R, –Cl and –OR, the Tl attacks almost exclusively the *para* position. However, for groups such
as $-\overset{\overset{O}{\|}}{C}-OR$, $-CH_2 OR$ and $-\overset{\overset{O}{\|}}{C}-OH$, the position of attack
is *ortho* (this is believed to be due to complexation of
the Tl with the substituent). Another important feature
of Tl substitution is that at higher temperatures, $> 70°C$,
the *meta* isomer is favored over the *ortho* or *para* isomer,

e.g., consider the above example, if the reaction is performed at 75°C, the major product would be

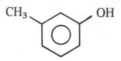

XIII-2 Cleavage of an aromatic ether with HI

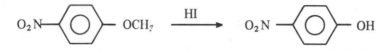

A third preparation of phenols involves the hydrolysis of a diazonium salt which is procedure XII-5.

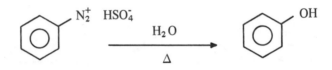

WORKED SYNTHESES

Starting with any monosubstituted cyclic molecule, any monosubstituted acyclic organic molecule containing no more than four carbons, and any inorganic reagents, prepare the following:

XIII-a

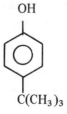

Let us ask some questions concerning XIII-a.

(1) What kind of molecule is XIII-a?
    Answer: IIII-a is *p*-substituted phenol (*p-tert*-butylphenol).

(2) What procedures are available for preparing a phenol?
Answer: Procedure XIII-1 involving a thallium derivative and procedure XIII-5 which involves the use of a diazonium salt. Since with alkyl groups the thallium reaction goes exclusively *para*, let us use procedure XIII-1. This will require the use of *t*-butylbenzene which is a permitted starting material.

Working the synthesis backwards from the desired product, the preparation of XIII-a becomes

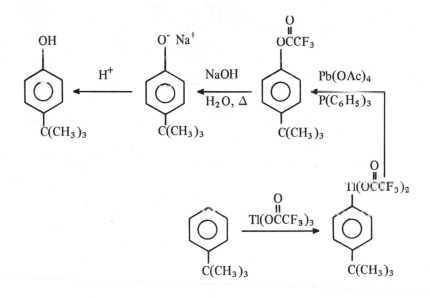

XIII-b

Let us ask some questions concerning XIII-b.

(1) What kind of molecule is XIII-b?
Answer: XIII-b is an *ortho* substituted phenol (*o*-bromophenol).

(2) What procedures are available for preparing phenols?
Answer: Procedure XIII-1 utilizing a thallium derivative
and procedure XIII-5 which requires a diazonium salt.

(3) How do we begin the synthesis of XIII-b?
Answer: The *ortho* substitution pattern of XIII-b is the
important feature of the synthetic scheme. The halogens
normally undergo the thallation reaction in the *para* position
at room temperature and in the *meta* position at elevated
temperatures, this tends to rule out the use of procedure
XIII-1. This leaves us with procedure XII-5.   This procedure
would require compound **A** to yield XIII-b

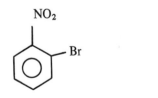

A

(4) How can **A** be prepared?
Answer: By diazotization of **B** via procedure XII-1.

B

(5) How can **B** be prepared?
Answer: By reduction of **C** via procedure XI-1.

C

(6) How can **C** be prepared?
Answer: Nitration in the *ortho* position can be accomplished
by using acetyl nitrate. Thus reacting bromobenzene via
procedure V-7 will yield **C**.

The synthesis of X-II-b becomes

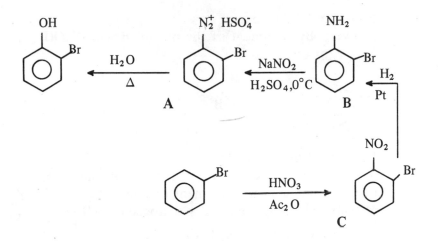

A

B

C

XIII-c

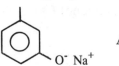

Let us ask some questions concerning XIII-c.

(1) What kind of molecule is XIII-c?
Answer: XIII-c is a *meta* disubstituted aromatic ring (*m*-butoxyphenetole).

(2) What procedures are available for placing an ether linkage on an aromatic ring?
Answer: Procedure XII-6 which requires a diazonium salt and procedure VII-1, which is the Williamson synthesis. Procedure VII-1 when applied to aromatic alkyl systems requires the sodium salt of a phenol and the appropriate alkyl halide. Since we are concerned with the synthesis of phenols in this chapter let us use procedure VII-1 which requires the phenoxide salt **A** and *n*-butyl bromide.

OCH₂CH₃

A

O⁻ Na⁺

(3) How can **A** be prepared?
    Answer: By treatment of the phenol **B** with NaOH.

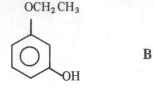

B

(4) How can **B** be prepared?
    Answer: Since **B** is a *meta* substituted phenol, it can most easily be prepared by treatment of phenetole with $Tl(OC-CF_3)_3$ at temperatures above 75°C via procedure XIII-1.

    Thus the synthesis of XIII-c becomes

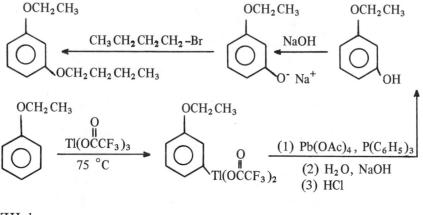

XIII-d

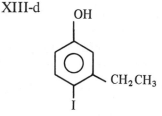

    Let us ask some questions concerning XIII-d.

(1) What kind of molecule is XIII-d?
    Answer: XIII-d is a disubstituted phenol (3-ethyl-4-iodophenol).

(2) What procedures are available for placing a phenol function on an aromatic ring?
Answer: Procedures XIII-1 involving thallation and procedure XII-5 involving a diazonium salt.

(3) What procedures are available for placing an ethyl group on an aromatic ring?
Answer: Procedure V-1 which is the Friedel-Crafts alkylation reaction and procedure X-6 which is the Friedel-Crafts acylation followed by a Wolff-Kischner reduction.

(4) What procedure is available for placing an I on an aromatic ring?
Answer: Procedure V-5 which involves the use of the I^+ electrophile.

(5) How do we start the synthesis of XIII-d?
Answer: There are two items which must be taken into account before the synthesis of XIII-d is commenced.
(1) The *meta* arrangement of the ethyl and hydroxy groups, each of which is an *ortho para* director. However, as stated in procedure XIII-1, the thallation of an alkyl substituted aromatic ring at high temperatures will yield a *meta* substituted phenol. (2) Since we are working the synthesis backwards, which group should be placed on the ring last? Because we have a procedure which will permit the *meta* orientation of the –OH and –C_2H_5 groups, the I should be placed on the ring in the last step. Another reason for placing the I on the ring in the last step is that we have fewer procedures for incorporation I into the molecule than for –C_2H_5 or –OH. Thus, we need compound A.

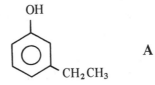

OH

CH$_2$CH$_3$

A

(6) How can A be prepared?
Answer: By thallation of ethyl benzene at 75°C using procedure XIII-1.

Thus the synthesis of XIII-d becomes

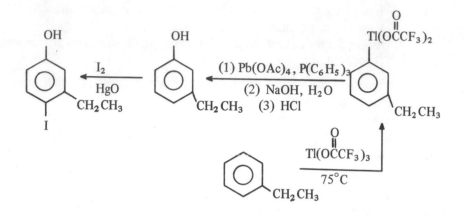

XIII-e

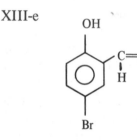

Let us ask some questions concerning XIII-e.

(1) What kind of molecule is XIII-e?
Answer: XIII-e is a disubstituted benzaldehyde (2-hydroxy-5-bromobenzaldehyde).

(2) What procedures are available for placing an –OH group on an aromatic ring?
Answer: Procedure XIII-1 and procedure XII-5.

(3) What procedures are available for placing a Br on an aromatic ring?
Answer: Procedure V-4 which involves the electrophilic substitution on an aromatic ring and procedure XII-3 which requires a diazonium salt.

(4) What procedures are available for placing a $-\overset{\overset{\text{O}}{\|}}{\text{C}}$–H group on an aromatic ring?

Answer: Procedure X-3 which is the oxidation of a methyl group attached to an aromatic ring. A procedure for placing an aldehyde function *ortho* to an –OH group on an aromatic ring is the Reimer-Tiemann reaction. It proceeds as illustrated below:

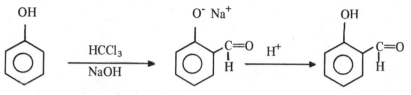

(5) How should we begin the synthesis of XIII-c?
Answer: It can be easily discerned that the *ortho* arrangement of the –OH and –CHO groups in XIII-e lends itself very readily to the Reimer-Tiemann reaction. If we desire to place the formyl group on the ring in the last step, this would require compound **A**.

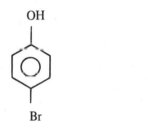

(6) How can **A** be prepared?
Answer: By bromination of phenol. However, since phenol is such a reactive molecule towards electrophilic substitution, the solvent used is CS_2 and the Lewis acid catalyst is omitted. This modification of procedure V-5 prevents polybromination of phenol. Thus the synthesis of XIII-e becomes

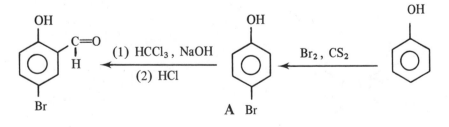

UNWORKED SYNTHESES

Using the same instructions stipulated before the worked syntheses section, prepare the following:

(f)
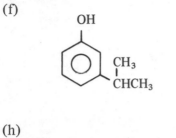

(g)
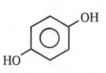

(h)
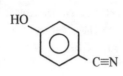

(i)

(j)
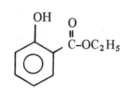

(k)

(l)
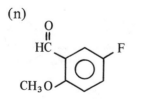

(m)
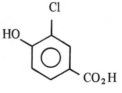

(n)

(o)

(p)

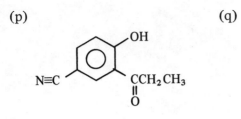

(q)

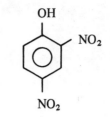

CHAPTER XIV

CONDENSATION REACTIONS

The number of condensation reactions which have been reported are too vast to be listed in this book. Thus, we shall confine ourselves to syntheses involving some of the more common condensations. The reader is referred to an organic text for mechanistic and more detailed experimental data.

CONDENSATION PROCEDURES

XIV-1 The aldol condensation
   The reaction is concerned with the self-condensation of aldehydes and ketones under acidic or basic conditions.

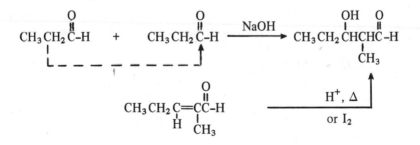

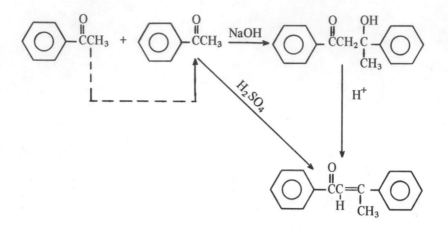

Important points regarding the aldol condensation.

(1) The use of aqueous NaOH results in the isolation of a β-hydroxy ketone provided the alcohol is not a substituted benzyl alcohol. The alcohol can be heated with acid or I_2 to yield the α,β-unsaturated carbonyl molecule. In cases where the initially formed alcohol is a derivative of benzyl alcohol, a molecule of H_2O is lost spontaneously.

(2) The use of H_2SO_4 or HCl results in the isolation of the α,β-unsaturated carbonyl compound directly.

(3) Aldehydes and ketones which do not contain an α-hydrogen do not undergo the aldol condensation.

XIV-2 Mixed aldol condensation

This reaction is concerned with the condensation of aldehydes and ketones of different structure. This reaction works best if one of the carbonyl molecules contains no α-hydrogens.

Base promoted

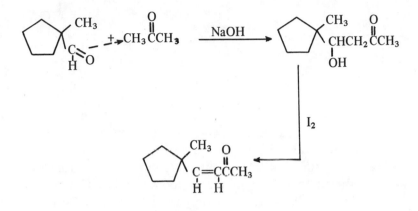

Acid catalyzed

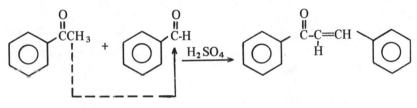

Note that once again for non-benzyl alcohol derivatives the intermediate β-hydroxy carbonyl compound can be isolated under base conditions. Acid conditions will yield the α,β-unsaturated carbonyl derivative directly.

XIV-3 Aldol-type condensations of nitro compounds

Here the α-hydrogen of the nitro molecule is more acidic than the α-hydrogen of an aldehyde or ketone. The carbanion formed from the nitro compound will then react with the C of an aldehyde or ketone
$$\begin{matrix} \text{C} \\ \| \\ \text{O} \end{matrix}$$

$$\underset{\text{O}}{\overset{\overset{\text{O}}{\|}}{\text{CH}_3\text{CH}_2\text{C-H}}} + \text{CH}_3\text{CH}_2\text{-NO}_2 \xrightarrow{\text{KOH}} \underset{\text{NO}_2}{\overset{\text{OH}}{\text{CH}_3\text{CHCHCH}_2\text{CH}_3}} \xrightarrow[\text{Pt}]{\text{H}_2}$$

$$\underset{\text{NH}_2}{\overset{\text{OH}}{\text{CH}_3\text{CHCHCH}_2\text{CH}_3}}$$

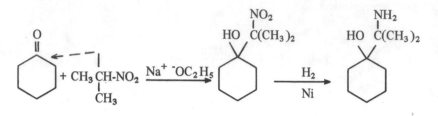

Important features concerning the above reaction.

(1) The base frequently used is KOH; however, other strong bases may also be used.

(2) The initially isolated β-nitro alcohol can be catalytically reduced to yield a β-amino alcohol. Thus this synthesis is an excellent path for preparing the aforementioned compounds.

XIV-4 The Claisen condensation

This reaction involves the self-condensation of an ester. The reaction is normally base catalyzed. The bases used are Na^+ $^-OC_2H_5$ (NaOEt), NaH and Na^+ $^-C(C_6H_5)_3$ [$NaC\phi_3$].

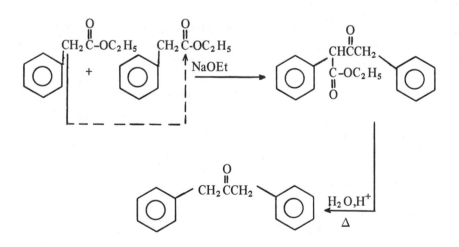

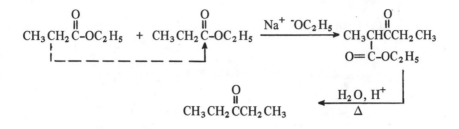

Important points regarding the Claisen condensation.

(1) If one uses NaOEt as the base then there must be two α-hydrogens for the reaction to go. If there is only one α-hydrogen, either NaH or NaCϕ₃ must be used as the base (see below).

(2) The initial β-keto ester can be isolated or it can be hydrolyzed and decarboxylated to yield a symmetrical ketone.

For esters which contain one α-hydrogen:

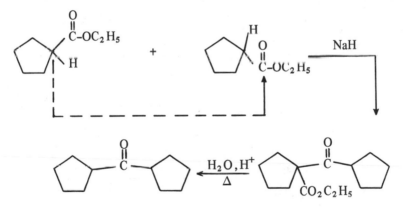

As noted above, Na^+ $^-C(C_6H_5)_3$ may also be used as the base.

XIV-5 Crossed Claisen condensation

This is a condensation between two different esters. Once again the base is NaOEt, NaH, or NaCϕ₃. This reaction works best if one of the esters has no α-hydrogens, *e.g.*,

(a) Ethyl formate

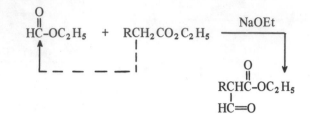

(b) Diethyl carbonate

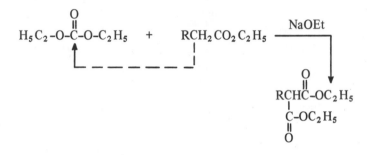

(c) Diethyl oxalate

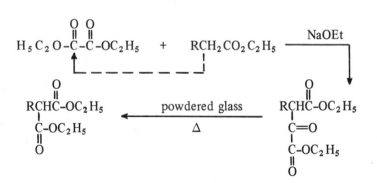

The last two syntheses yield derivatives of diethyl malonate, *e.g.*,

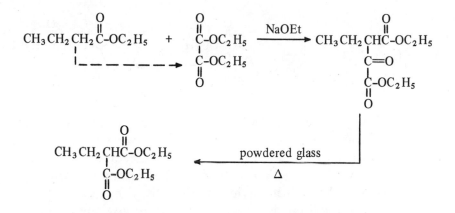

XIV-6 The condensation of esters with aldehydes and ketones

(a) Aldehydes: the $\overset{O}{\overset{\|}{C}}$ of the aldehyde group is condensed with the α-carbon of the ester. The bases used are NaOEt, NaH and $NaNH_2$.

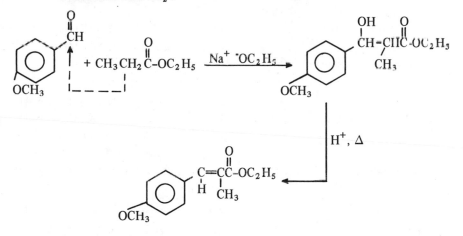

The initial product is a β-hydroxy ester, which may be dehydrated to an α,β-unsaturated ester.

(b) Ketone: the α-carbon of the ketone condenses with the C=O of the ester. The bases used are NaOEt, NaH and $NaNH_2$.

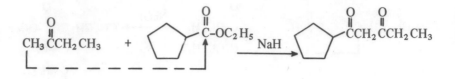

The product is a β-diketone. Note that in the case of the ketone there are two different α-hydrogens which may be removed. Because of the greater acidity and steric accessibility of the methyl hydrogens, they are removed more easily than the methylene hydrogens. Thus the methyl carbon will condense with the ester to give the major product.

Data Summary Regarding the Condensation Reactions in Chapter XIV

Desired Product	Procedure	Starting Material	Acid or Base
β-hydroxy aldehyde	XIV-1	two aldehydes	NaOH
α,β-unsaturated aldehyde	XIV-1	two aldehydes	$NaOH + I_2$
β-keto alcohol	XIV-2	two ketones	NaOH
α,β-unsaturated ketone	XIV-2	two ketones	$NaOH + I_2$ or $H^+ + \Delta$
β-hydroxy ester	XIV-6a	aldehyde + ester	$NaOC_2H_5$ or NaH
α,β-unsaturated ester	XIV-6a	aldehyde + ester	$NaOC_2H_5$ or NaH $+ I_2$ or $NaOC_2H_5 + H^+ + \Delta$
α-formyl ester	XIV-5a	ester + ethyl formate	$NaOC_2H_5$ or NaH
β-keto ester	XIV-4	two esters	$NaOC_2H_5$ or NaH
β-diketone	XIV-6b	ketone + ester	$NaOC_2H_5$, NaH or $NaNH_2$
β-nitro alcohol	XIV-3	aldehyde + nitroalkane ketone + nitroalkane	KOH or NaH
malonic esters	XIV-5b	ester + diethyl carbonate	KOH or NaH
		ester + diethyl oxalate	KOH or NaH

WORKED SYNTHESES

Using the procedures in Chapter XIV prepare the following compounds.

XIV-a

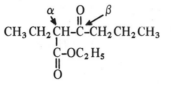

Let us ask some questions concerning XIV-a.

(1) What kind of molecule is XIV-a?
Answer: XIV-a is a β–keto ester (ethyl 2-ethyl-3-keto-hexanoate).

(2) What procedure is available for preparing β-keto esters?
Answer: Procedure XIV-4 which requires condensing two esters.

(3) How can XIV-a be broken up so that the structures of the two esters required for this synthesis can be discerned?
Answer: The α,β-bond must be cleaved as illustrated below. Looking at procedure XIV-4 it is obvious that the ketone function in the product was an ester carbonyl in the starting material. Thus, we must add an –OR (–OC$_2$H$_5$) to the carbonyl to get the structure of the ester which will yield the ketone in the product. To the carbon α to the ester function we must add one H.

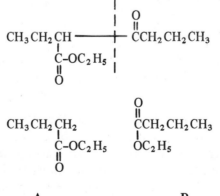

A   B

It can be easily seen that **A** and **B** are the same. Thus XIV-a can be synthesized by the self-condensation of ethyl butyrate.

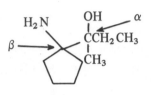

$$+ \quad CH_3CH_2CH_2CO_2C_2H_5$$

XIV-b

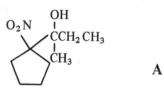

Let us ask some questions concerning XIV-b

(1) What kind of molecule is XIV-b?
Answer: XIV-b is a β-aminoalcohol [2-(1-aminocyclopentyl)-2-butanol].

(2) What procedure is available for preparing β-amino alcohols?
Answer: Procedure XIV-3 which involves the catalytic reduction of a β-nitro alcohol. Thus we need **A**.

$$\begin{array}{c} OH \\ O_2N \quad | \\ CCH_2CH_3 \\ | \\ CH_3 \end{array}$$

A

(3) What procedure is available for preparing β nitro alcohols?
Answer: Procedure XIV-3 which involves a condensation between an aliphatic nitro compound and an aldehyde or ketone.

(4) How can **A** be broken up so that the structures of the required starting materials can be discerned?
Answer: To find the needed starting materials, the bond between the α- and β-carbons must be cleaved as shown below. The carbinol carbon in the product must be the carbonyl carbon in the starting aldehyde or ketone as is shown in procedure XIV-3.

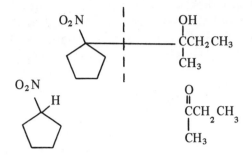

Thus the synthesis of XIV-b becomes

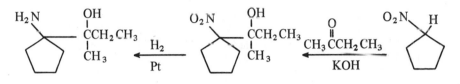

XIV-c

Let us ask some questions concerning XIV-c.

(1) What kind of molecule is XIV-c?
Answer: XIV-c is a β-diketone (1-phenyl-4-methyl-1,3-pentadione).

(2) What procedure is available for preparing β-diketones?
Answer: Procedure XIV-6b which requires a ketone and an ester.

(3) How can XIV-c be broken up so that the required starting ketone and ester can be deduced?
Answer: To find the structures of the required starting materials the carbon situated between the two carbonyl groups must suffer cleavage of one of its carbon carbonyl-carbon bonds as illustrated below. Note that since there

are two such bonds there are two ways of cleaving the mole-cule. To reconstruct the required ester needed in the pro-cedure, an –OR is added to the carbonyl, which suffered bond cleavage. This yields the starting materials **A, B, C** and **D.**

Cleavage 1   Cleavage 2

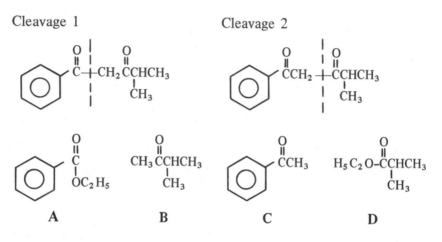

A   **B**   **C**   **D**

Using compounds **A** and **B** the synthesis of XIV-c becomes

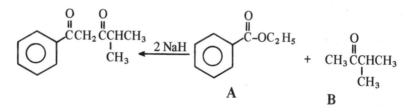

A   **B**

XIV-d

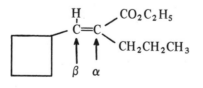

Let us ask some questions concerning XIV-d.

(1) What kind of molecule is XIV-d?
   Answer: XIV-d is an α,β-unsaturated ester (ethyl 2-*n*-propyl-3-cyclobutyl-3-propenoate).

(2) What procedure is available for preparing α,β-unsaturated esters?

Answer: Procedure XIV-6a which involves the condensation of an aldehyde and an ester followed by the elimination of H_2O.

(3) How can XIV-d be broken up so that the structures of the required starting materials can be discerned?

Answer: The olefinic double bond must first be cleaved. Looking at procedure XIV-6a it can readily be seen that the β-carbon in XIV-d was the aldehyde carbon in the starting material. We thus convert the cleavage product containing the β-carbon into the aldehyde **A**. The α-carbon of XIV-d comes from the starting ester. We thus convert that cleavage product to an ester by adding H.

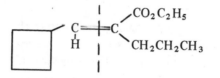

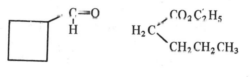

A   B

Thus the synthesis of XIV-d becomes

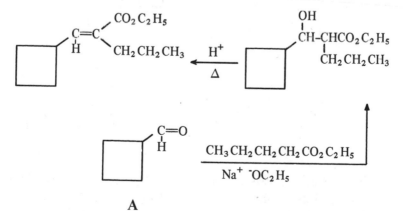

A

XIV-e

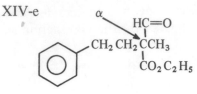

Let us ask some questions concerning XIV-e.

(1) What kind of molecule is XIV-e?
Answer: XIV-e is an α-formyl ester (ethyl 2-methyl-4-phenyl-2-formylbutanoate).

(2) What procedure is available for preparing α-formyl esters?
Answer: Procedure XIV-5a which requires an ester and ethyl formate.

(3) How can XIV-e be broken up so that the structures of the desired starting materials can be discerned?
Answer: The carbon aldehyde-carbon bond must be cleaved. Looking at procedure XIV-5a, it can be readily seen that the formyl group in XIV-e comes from ethyl formate **B** while the other carbon cleaved must come from the starting ester **A**. This carbon must have had a hydrogen removed in the formation of the product XIV-e. Thus we must reattach a hydrogen to identify the structure of the starting ester.

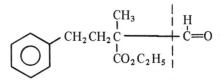

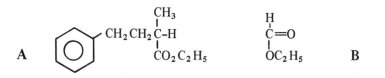

Note that **A** possesses only one α-hydrogen. Thus we must use NaH as the base to get a better yield.
The synthesis of XIV-e becomes

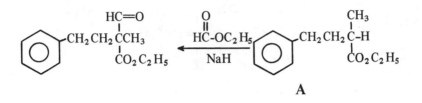

A

XIV-f

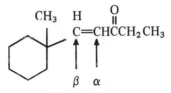

Let us ask some questions concerning XIV-f.

(1) What kind of molecule is XIV-f?
Answer: XIV-f is an α,β-unsaturated ketone [1-(1-methyl-cyclohexyl)-1-penten-3-one].

(2) What procedure is available for preparing α,β-unsaturated ketones?
Answer: Procedure XIV-2 which involves the condensation of an aldehyde and a ketone followed by the elimination of a molecule of H_2O.

(3) How can XIV-f be broken up so that the structures of the required starting materials can be deduced?
Answer: The double bond between the α- and β-carbons must be cleaved. Looking at procedure XIV-2 one can readily ascertain that the α-carbon in XIV-f comes from the ketone while the β-carbon is derived from the aldehyde. We can now reconstruct the starting materials by cleaving the double bond and adding a H to the α-carbon and converting the β-carbon to an aldehyde group as shown below.

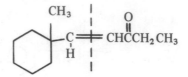

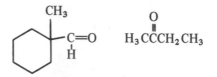

Thus the synthesis of XIV-f becomes

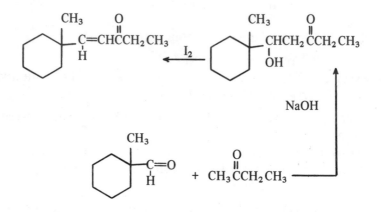

NaOH

XIV-g
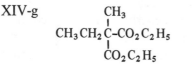

$$CH_3CH_2\underset{\underset{CO_2C_2H_5}{|}}{\overset{\overset{CH_3}{|}}{C}}-CO_2C_2H_5$$

Let us ask some questions concerning XIV-g.

(1) What kind of molecule is XIV-g?
Answer: XIV-g is a disubstituted malonic ester (diethyl methylethylmalonate).

(2) What procedures are available for preparing substituted malonic esters?
Answer: Procedure XIV-5b which requires an ester and diethyl carbonate and procedure XIV-5c which requires an ester and diethyl oxalate. Let us use procedure XIV-5b.

(3) How can XIV-g be broken up so that the structures of the required starting materials can be discerned?
Answer: We must cleave one of the carbon $-CO_2C_2H_5$ bonds. Looking at procedure XIV-g it is readily apparent that one of the $-CO_2C_2H_5$ group was derived from diethyl carbonate. To reconstruct that molecule we merely add $-OC_2H_5$ to the $-CO_2C_2H_5$ group we cleaved. The carbon from which the $-CO_2C_2H_5$ group was severed must be part of the starting ester required for the synthesis; to this carbon we add H.

Thus the synthesis of XIV-g becomes

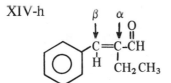

It is necessary to use NaH because the starting ester has only one α-hydrogen.

XIV-h

$$\underset{\underset{H}{\overset{\beta}{\downarrow}}}{}\ \underset{\underset{CH_2CH_3}{\overset{\alpha}{\downarrow}}}{C=C-CH}\overset{O}{\overset{\|}{}}$$

Let us ask some questions concerning XIV-h.

(1) What kind of molecule is XIV-h?
Answer: XIV-h is an α,β-unsaturated aldehyde (α-ethylcinnamaldehyde).

(2) What procedure is available for preparing an α,β-unsaturated aldehyde?
Answer: Procedure XIV-1 which involves the condensation of two aldehydes under acidic or basic conditions.

(3) How can XIV-h be broken up so that the structures of the required starting materials can be discerned?
Answer: The α,β-double bond must be cleaved. From procedure XIV-1 it can be seen that the β-carbon in XIV-h was an aldehyde carbon in the starting material **A**. The α-carbon

of XIV-h must contain two hydrogens. If we add two hydrogens, we have the other starting material **B**.

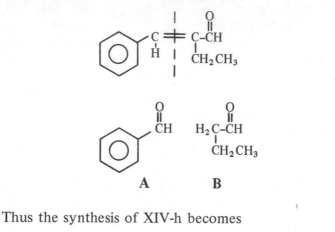

A   B

Thus the synthesis of XIV-h becomes

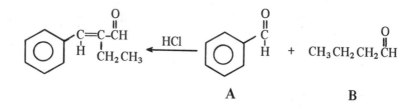

A   B

UNWORKED SYNTHESES

Synthesize the following using the instructions before the worked syntheses section.

(i)

$$CH_2CH_3$$
$$C\text{-}CO_2C_2H_5$$
$$CO_2C_2H_5$$

(j)

$$OH \quad NH_2$$
$$CH_3C\text{---}CCH_2CH_3$$
$$CH_3 \quad CH_3$$

(k)

$$CCH_2CH_3$$

(l)

$$CH_3 \quad O \quad CO_2C_2H_5$$
$$CH_3C\text{---}C\text{-}CCH_3$$
$$H \qquad CH_3$$

(m)

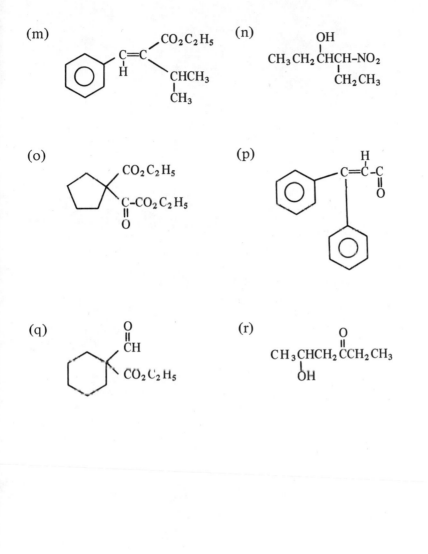

$CO_2C_2H_5$

$C=C$

H

$CHCH_3$

CH_3

(n)

OH

$CH_3CH_2CHCH-NO_2$

CH_2CH_3

(o)

$CO_2C_2H_5$

$C-CO_2C_2H_5$

O

(p)

H

$C=C-C$

O

(q)

O

CH

$CO_2C_2H_5$

(r)

O

$CH_3CHCH_2CCH_2CH_3$

OH

CHAPTER XV

OTHER CARBON–CARBON BOND FORMING REACTIONS

PROCEDURES

XV-1   Acetoacetic ester synthesis

Important points regarding the acetoacetic ester synthesis.

(1)   The reaction proceeds via a carbanion. Thus, the reaction works only if the alkyl halide is primary or secondary; tertiary halides and aromatic halides will not work.

(2)   The initial product is a β-keto ester which may be hydrolyzed with acid to a β-keto acid; the acid may be decarboxylated using heat to yield a methyl ketone. Thus, if one desires to prepare a β-keto ester, a β-keto acid (structure must contain a $CH_3\overset{O}{\overset{\|}{C}}$ grouping) or a

methyl ketone, the acetoacetic ester synthesis is a possible route.

XV-2   Malonic ester synthesis

$$H_2C(CO_2C_2H_5)_2 \xrightarrow{\text{NaOEt}} \overset{\text{Na}^+}{H\bar{C}(CO_2C_2H_5)_2} \xrightarrow{RX} \overset{R}{\underset{}{H\overset{|}{C}(CO_2C_2H_5)_2}}$$

$$\downarrow \text{Na}^+ \ ^-OC_2H_5$$

$$\underset{\overset{|}{CO_2H}}{\overset{R'}{\underset{}{R\overset{|}{C}-CO_2H}}} \xleftarrow[H^+]{H_2O} \underset{}{\overset{R'}{\underset{}{R\overset{|}{C}(CO_2C_2H_5)_2}}} \xleftarrow{R'X} \underset{\text{Na}^+}{R\underline{C}(CO_2C_2H_5)_2}$$

$$\downarrow \underset{-CO_2}{\Delta} \quad \underset{}{\overset{R'}{\underset{}{R\overset{|}{C}HCO_2H}}}$$

Important points regarding the malonic ester synthesis.

(1) Like the acetoacetic ester reaction, the malonic ester synthesis also proceeds via a carbanion and thus the alkyl halide should be either primary or secondary. Tertiary and aromatic halides will not work.

(2) The initial product from the malonic ester ester synthesis is a derivative of malonic ester, which may be hydrolyzed to a derivative of malonic acid. Decarboxylation readily occurs to yield a carboxylic acid.

(3) The malonic ester can be used to form cyclic molecules. The other material needed is an ω, ω' dihalide.

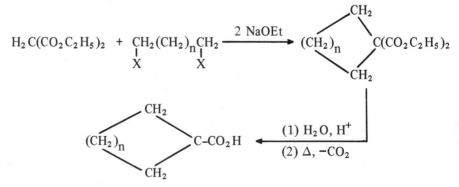

XV-3   The condensation of nitriles with aldehydes and ketones.

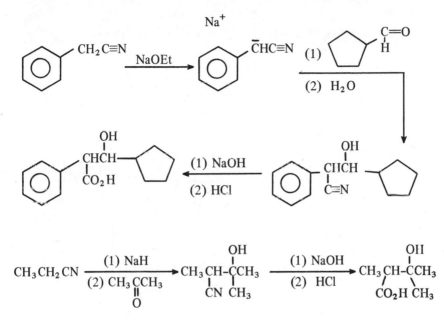

Note: This is an excellent method of preparing either β hydroxy nitriles or β-hydroxy acids. If the alcohol desired is a secondary alcohol, an aldehyde should be condensed with the nitrile. If the alcohol desired is a tertiary alcohol, a ketone should be condensed with a nitrile.

XV-4   Reaction of organoboranes.
       This reaction is used to prepare

Ketones

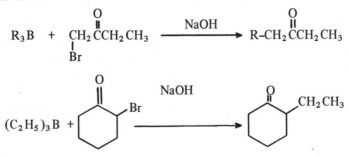

Esters and Acids

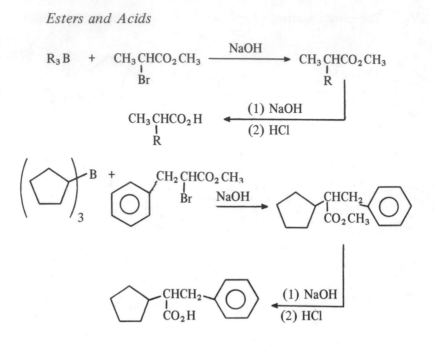

Note: This reaction requires either an α-halo ketone or an α-halo ester.

XV-5   Stork-enamine reaction
This reaction can be used to prepare either a β-diketone or a β-keto aldehyde. Synthesis of the enamine requires an aldehyde or ketone and a secondary amine (frequently pyrrolidine   ).

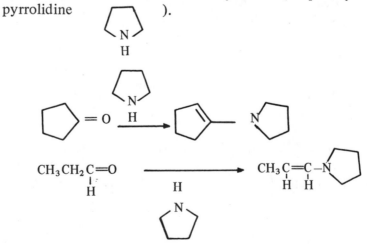

To prepare a β-diketone, we must start with the enamine of a ketone and treat it with an acid chloride.

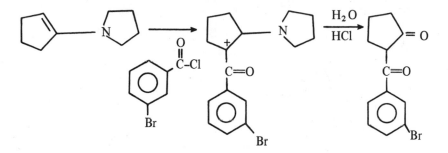

This reaction works best if the starting ketone is symmetrical. To prepare a β-keto aldehyde, we must start with the enamine of an aldehyde and treat it with an acid chloride.

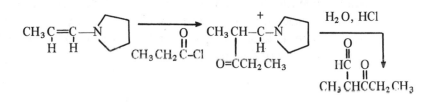

DATA SUMMARY REGARDING THE REACTIONS IN CHAPTER XV

Desired Product	Procedure	Starting Materials
methyl ketone	XV-1	acetoacetic ester + alkyl halide
β-keto acid	XV-1	acetoacetic ester + alkyl halide
substituted malonic acid	XV-2	malonic ester + alkyl halide
carboxylic acid	XV-2	malonic ester + alkyl halide
β-hydroxy nitriles	XV-3	alkyl nitrile + aldehyde or ketones
β-hydroxy acids	XV-3	alkyl nitrile + aldehyde or ketone
α-alkylated ketone	XV-4	alkyl borane + α-halo ketone
α-alkylated ester	XV-4	alkyl borane + α-halo ester
β-diketone	XV-5	enamine + acid chloride
β-keto aldehyde	XV-5	enamine + acid chloride

WORKED SYNTHESES

Using the procedures in Chapter XV, prepare the following compounds.

XV-a

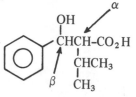

Let us ask some questions concerning XV-a.

(1) What kind of molecule is XV-a?
Answer: XV-a is a β-hydroxy acid (2-isopropyl-3-phenyl-3-hydroxypropanoic acid).

(2) What procedure is available for preparing β-hydroxy acids?
Answer: Procedure XV-3 which requires an alkyl nitrile and an aldehyde if the alcohol is secondary, followed by hydrolysis. Thus we need compound **A**.

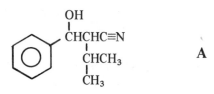

(3) How can **A** be broken up so that the structures of the required starting materials can be discerned?
Answer: Looking at procedure XV-3, it can be readily determined that the bond formed during this synthesis is the α,β-bond of **A**. To determine the starting materials needed, the α,β-bond must be cleaved. The carbinol carbon becomes a carbonyl carbon in the starting materials. The α-carbon of **A** is the carbon adjacent to the –CN group in the starting nitrile.

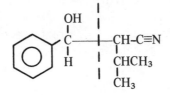

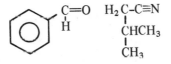

Thus the synthesis of XV-a becomes

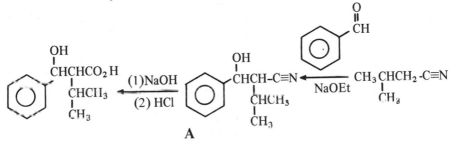

A

XV-b

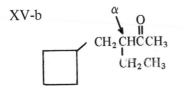

Let us ask some questions concerning XV-b.

(1) What kind of molecule is XV-b?
Answer: XV-b is an aliphatic methyl ketone (3-ethyl-4-cyclopentyl-2-butanone).

(2) What procedure is available in Chapter XV for preparing aliphatic methyl ketones?
Answer: Procedure XV-1 which is the acetoacetic ester synthesis.

(3) How do we begin the synthesis of XV-b using procedure XV-1?

Answer: The first step is to recreate the acetoacetic ester molecule. The last step in procedure XV-1 in which a ketone is produced involves a decarboxylation in which the $-CO_2H$ group on the carbon α to the carbonyl group is replaced by a H. To reverse this step we take the H on the α-carbon in XV-b and replace it with a $-CO_2H$ group yielding **A**.

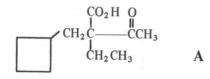

(4) How can **A** be prepared?

Answer: Looking at procedure XV-1, we see that the acid is derived from the corresponding ethyl ester. Thus, we need **B**.

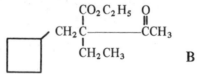

(5) How can **B** be prepared?

Answer: By treating **C** with ICH_2CH_3

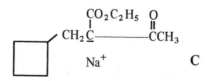

(6) How can **C** be prepared?

Answer: By treating **D** with $NaOC_2H_5$.

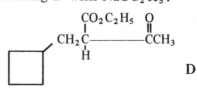

(7) How can **D** be prepared?
Answer: By treating **E** with ICH_2

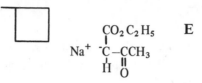

(8) How can **E** be prepared?
Answer: By treating acetoacetic ester with $NaOC_2H_5$.
Thus the synthesis of XV-b becomes

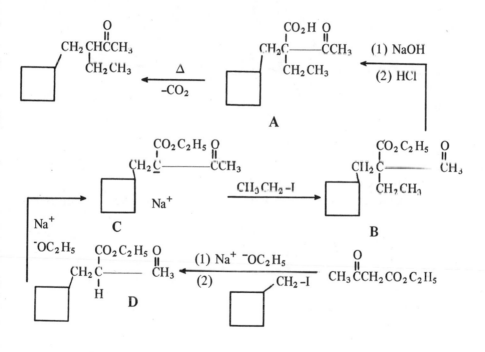

XV-c

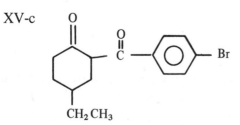

Let us ask some questions concerning XV-c.

(1) What kind of molecule is XV-c?
Answer: XV-c is a β-diketone. [4-ethyl-2-(4-bromobenzoyl)
cyclohexanone].

(2) What procedure is available for preparing β-diketones?
Answer: Procedure XV-5 which requires an enamine and an
acid chloride.

(3) Which carbonyl of XV-c should be converted into an enamine?
Answer: The cyclohexanone carbonyl is by far the easiest.
Thus, we need **A**, which upon hydrolysis will yield XV-c.

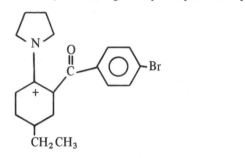

A

(4) How can **A** be prepared?
Answer: By treating **B** with the acid chloride 4-bromobenzoyl
chloride.

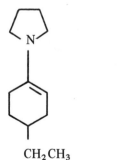

B

(5) How can **B** be prepared?
Answer: By treating 4-ethylcyclohexanone with pyrrolidine.
Thus, the synthesis of XV-c becomes

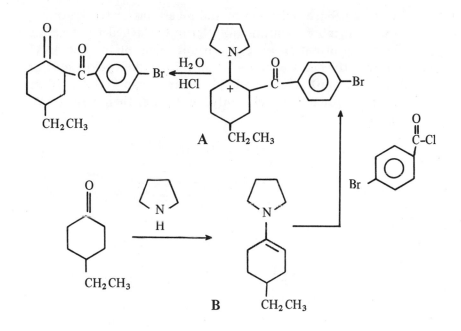

A   CH₂CH₃

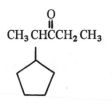

B   CH₂CH₃

XV-d

$$CH_3\ \underset{\underset{\displaystyle \text{(cyclopentyl)}}{|}}{CH}\overset{\displaystyle O}{\overset{\|}{C}}CH_2\ CH_3$$

Let us ask some questions concerning XV-d.

(1) What kind of molecule is XV-d?
Answer: XV-d is α-substituted aliphatic ketone (2-cyclo-pentyl-3-pentanone).

(2) What procedure in Chapter XV is available for preparing an α-alkylated non-methyl ketone?
Answer: Since XV-d is not a methyl ketone only procedure XV-4 which requires an α-halo ketone and a trialkyl borane is available.

(3) How can XV-d be broken up so that the structures of the required starting materials can be determined?

Answer: Since only one of the α-carbons of the ketone XV-d has more than one alkyl group attached to it we must focus our attention on that carbon. One of the two alkyl groups attached to the α-carbon (either the cyclopentyl or methyl group) must be added to the α-halo ketone. If we desire to place the cyclopentyl group on, then we need **A** and **B**.

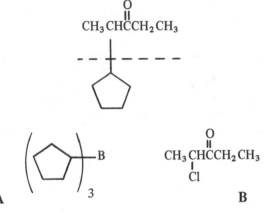

(4) How can **A** be prepared?
Answer: By treating 3-pentanone with Cl_2 and H_2O.

(5) How can **B** be prepared?
Answer: By reacting cyclopentene with diborane.
Thus the synthesis of XV-d becomes

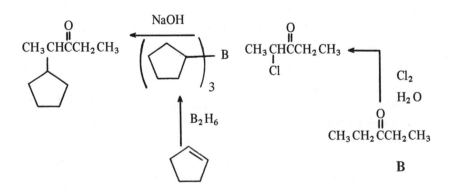

XV-e

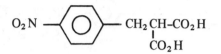

Let us ask some questions about XV-e.

(1) What kind of molecule is XV-e?
Answer: XV-e is a substituted malonic acid (4-nitrobenzyl-malonic acid).

(2) What procedure in Chapter XV is available for preparing substituted malonic acids?
Answer: Procedure XV-2 which requires the hydrolysis of a malonic ester. Thus, to prepare XV-e, we need **A**.

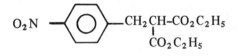

A

(3) How can **A** be prepared?
Answer: Malonic ester can be alkylated; thus we merely have to treat the Na salt of malonic ester **B** with the alkyl halide 4-nitrobenzyl bromide.

$$Na^+ \ ^-\underset{H}{C}(CO_2C_2H_5)_2$$

B

(4) How can **B** be prepared?
Answer: By treating malonic ester with $NaOC_2H_5$.
Thus the synthesis of XV-e becomes

$$O_2N-\!\!\!\bigcirc\!\!\!- CH_2CH\!-\!CO_2H \xleftarrow[HCl]{H_2O} O_2N-\!\!\!\bigcirc\!\!\!- CH_2CH\!-\!CO_2C_2H_5$$
$$\hspace{3cm} CO_2H \hspace{5cm} CO_2C_2H_5$$
$$\hspace{9.5cm} \textbf{A}$$

$$O_2N-\!\!\!\bigcirc\!\!\!- CH_2Br$$

$$H_2C(CO_2C_2H_5)_2 \xrightarrow{Na^+\ {}^-OC_2H_5} Na^+\ {}^-\underset{H}{C}(CO_2C_2H_5)_2$$
$$\hspace{9.5cm}\textbf{B}$$

XV-f

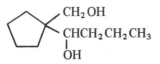

Let us ask some questions regarding XV-f.

(1) What kind of molecule is XV-f?
Answer: XV-f is a 1,3-diol [1-(1-hydroxymethylcyclopentyl) butanol].

(2) What procedures are available in Chapter XV for preparing 1,3-diols?
Answer: None. However, alcohols can be prepared by reducing carbonyl groups; thus reducing **A** would yield XV-f.

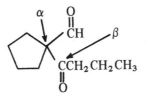

A

Note that **A** is a β-keto aldehyde, which can be prepared via procedure XV-5 which requires an enamine of an aldehyde and an acid chloride.

(3) How can **A** be prepared?
     Answer: By hydrolyzing **B**.

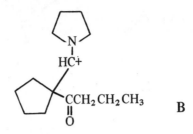

B

(4) How can **B** be prepared?
     Answer: By adding butanoyl chloride to enamine **C**.

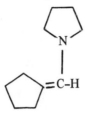

C

(5) How can **C** be prepared?
     Answer: By reacting **D** with pyrrolidine.

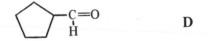

D

Thus the synthesis of XV-f becomes

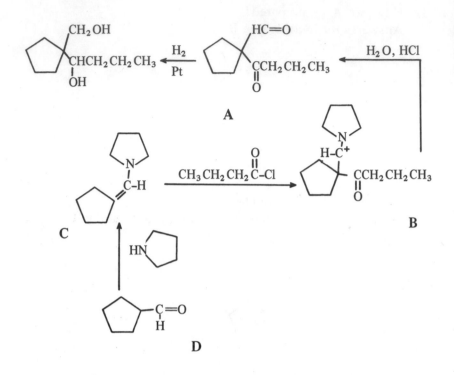

A

B

C

D

UNWORKED SYNTHESES

Using the instructions stated before the worked syntheses section, prepare the following compounds.

(g)

(h)

(i)

(j)

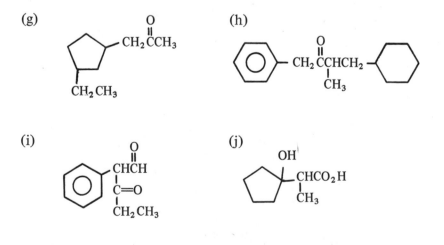

(k)

$$\begin{array}{ccc} CH_3 & OH & CH_3 \\ | & | & | \\ CH_3C\!\!-\!\!CH\!\!-\!\!CCH_2CH_3 \\ | & | \\ CH_3 & CO_2H \end{array}$$

(l)

$$\begin{array}{c} OH \\ | \\ CH_3CH_2CHCHCH_2CH_2CH_3 \\ CH_3CH_2CH\!\!-\!\!OH \end{array}$$

(m)

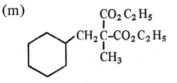

$$\begin{array}{c} CO_2C_2H_5 \\ | \\ CH_2C\!\!-\!\!CO_2C_2H_5 \\ | \\ CH_3 \end{array}$$

(n)

$$\begin{array}{c} CH_3CHCH_2CH_3 \\ | \\ HO_2C\!\!-\!\!C\!\!-\!\!CO_2H \end{array}$$

(o)

$$\begin{array}{cc} OH & CH_3 \\ | & | \\ CH_3CH_2CH\!\!-\!\!CCH_2\!\!-\!\!OH \\ & | \\ & CH_3 \end{array}$$

CHAPTER XVI

NAPHTHALENE DERIVATIVES

There are many reactions which allow one to synthesize molecules containing the naphthalene nucleus. However, for the purpose of brevity, we will limit this chapter to a discussion of the most frequently encountered reactions which are used to synthesize the most common derivatives of naphthalene. Once again the student should consult an organic text to get a fuller picture regarding the synthesis of naphthalenes.

In the following procedures reference will be made to the α, β, α′ and β′ positions of naphthalene; the following structure indicates these positions.

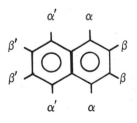

PROCEDURES FOR SYNTHESIZING
SUBSTITUTED NAPHTHALENES

XVI-1 Halogenation of naphthalene

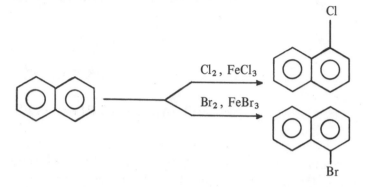

The halogen attacks the α-position to the tune of about 95%.

XVI-2 Nitration of naphthalene

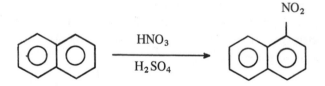

Nitration strongly favors the α-position over the β-position by about 19:1.

XVI-3 Sulfonation of naphthalene

Note that the position of attack for the –SO$_3$H group on the naphthalene nucleus is temperature dependent.

XVI-4 Friedel-Crafts acylation of naphthalene.

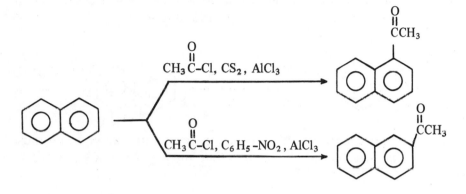

Note that the position of substitution is controlled by the solvent.

XVI-5 Preparation of naphthylamines
The easiest preparation of naphthylamines involves the catalytic reduction of the nitronaphthalenes.

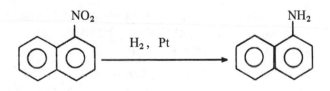

XVI-6 Preparation of naphthols
The naphthalene analogs of phenols can be placed on the naphthalene nucleus by hydrolyzing the appropriate diazonium salt.

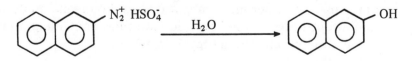

XVI-7 Preparation of alkyl naphthalenes

The Friedel-Crafts alkylation of naphthalene often yields a variety of products. For this reason it is not frequently used. A better method of placing an alkyl group involves reduction of the ketone formed from procedure XVI-4 using either the Wolff-Kischner or Clemmensen reductions.

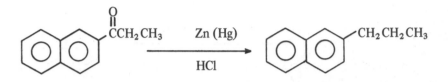

In order to use this procedure, the desired alkyl group must be attached to the naphthalene nucleus via a $-CH_2$ group.

DISUBSTITUTION OF NAPHTHALENE

There are three rules which control the position of electrophilic substitution attack on a monosubstituted naphthalene.

(1) If a deactivating group is present on a monosubstituted naphthalene, the incoming electrophile attacks the α' position of the other ring (sulfonation may go to the β' position).

(2) If an activating group is present in the 1-position on a monosubstituted naphthalene the incoming electrophile goes to the 4-position.

(3) If an activating group is present in the 2-position on a monosubstituted naphthalene, the incoming electrophile will go to the 1-position.

WORKED SYNTHESES

Starting with naphthalene prepare the following compounds. You may use any other monosubstituted cyclic organic molecules, any monosubstituted acyclic organic molecules containing no more than four carbons and any inorganic reagents.

XVI-a

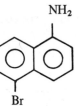

Let us ask some questions concerning XVI-a.

(1) What kind of molecule is XVI-a?
Answer: XVI-a is a disubstituted naphthalene (5-bromo-1-naphthylamine).

(2) What procedures are available for preparing naphthaylamines
Answer: Procedure XVI-5 which states that an amino group can be placed on a naphthalene nucleus by the catalytic reduction of the corresponding nitro compound (in this case compound **A**).

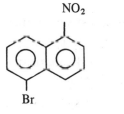

A

(3) How can **A** be prepared?
Answer: Since both the –Br and –NO₂ groups are deactivating toward electrophilic substitution, either 1-bromonaphthalene **B** or 1-nitronaphthalene **C** would place the required incoming electrophile in the correct position. Let us use **B** as our source of **A**. This will require the use of procedure XVI-2.

B

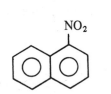

C

(4)   How can **B** be prepared?
Answer: By bromination of naphthalene via procedure XVI-1.
Thus, the synthesis of XVI-a becomes

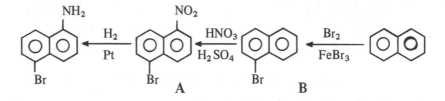

XVI-b

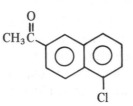

L Let us ask some questions concerning XVI-b.

(1)   What kind of molecule is XVI-b?
Answer: XVI-b is a disubstituted naphthalene [methyl-2-(5-chloronaphthyl)ketone].

(2)   How do we start the synthesis of XVI-b?
Answer: Both the Cl– and the $CH_3 \overset{O}{\underset{\|}{C}}$–group are deactivating.
Thus if we start with 1-chloronaphthalene and attempt to
acetylate it, the $CH_3 \overset{O}{\underset{\|}{C}}$– will go to one of the α' positions,
giving us the wrong product. If, however, we start with 2-acetylnaphthalene and chlorinate we will get XVI-b. Thus,
we need **A** and procedure XVI-1.

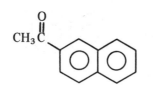

A

(3) How can **A** be prepared?
Answer: By reacting naphthalene with $CH_3\overset{\overset{\displaystyle O}{\|}}{C}Cl$ in ϕNO_2
via procedure XVI-2.
Thus the synthesis of XVI-b becomes

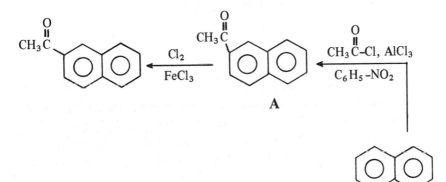

A

XVI-c

CH₂CH₂CH₃

Let us ask some questions concerning XVI-c.

(1) What kind of molecule is XVI-c
Answer: XVI-c is a dialkylated naphthalene (1,5-dipropyl-naphthalene).

(2) **What** procedure is available for placing an alkyl group on an aromatic ring?
Answer: The Friedel-Crafts alkylation reaction is useless in the naphthalene series because of the polyalkylation that frequently occurs. The best method of placing an alkyl group on a naphthalene nucleus involves the reduction of a ketone using the Wolff-Kischner or Clemmensen reductions as shown in procedure XVI-7. Thus we need **A**.

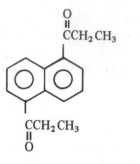

A

(3) How can **A** be prepared?

Answer: The acyl group is a deactivating group. Thus, if we start with 2-propanoylnaphthalene, we will isolate the wrong product 1,6-dipropanoylnaphthalene. If, however, we start with 1-propanoylnaphthalene **B** we should get **A**.

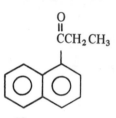

B

(4) How can **B** be prepared?

Answer: By treating naphthalene with propanoyl chloride in CS_2 via procedure XVI-4.

Thus the synthesis of XVI-c becomes

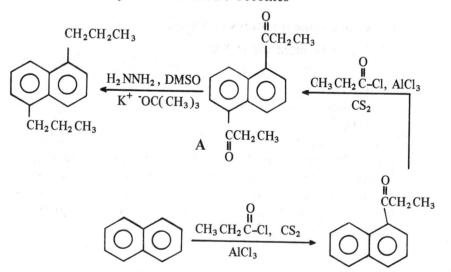

XVI-d

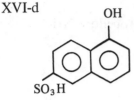

Let us ask some questions concerning XVI-d.

(1) What kind of molecule is XVI-d?
Answer: XVI-d is a hydroxyl substituted naphthalenesulfonic acid (5-hydroxy-2-naphthalenesulfonic acid).

(2) What procedure is available for placing an –OH group on a naphthalene nucleus?
Answer: Procedure XVI-6 which requires a diazonium salt and water.

(3) What procedure is available for placing a –SO_3H group on a naphthalene nucleus?
Answer: Procedure XVI-3, which requires H_2SO_4 and 160° to cause substitution in the β position.

(4) How should we begin the synthesis of XVI-d?
Answer: Note that the –OH is an activating group. If we sulfonated α-naphthol, the –SO_3H would attack either the 2- or 4-position depending upon the temperature used. The –SO_3H group is a deactivating group; thus if we nitrated 2-naphthalene-sulfonic acid, nitration would occur in the α'-position in the other ring. We could then reduce the nitro group to the amine. Diazotization and hydrolysis would then yield the correct product. Thus the last step in the synthesis should consist of conversion of **A** to XVI-d via procedure XVI-6.

N_2^+ HSO_4^-

A

SO_3H

(5) How can **A** be prepared?
Answer: By diazotization of **B** via procedure XII-1.

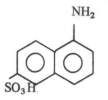

B

(6) How can **B** be prepared?
Answer: By reduction of **C** as shown in procedure XVI-5.

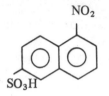

C

(7) How can **C** be prepared?
Answer: By nitrating **D** via procedure XVI-2.

D

(8) How can **D** be prepared?
Answer: By treating naphthalene with H_2SO_4 at 160°C
via procedure XVI-3. Thus the synthesis of XVI-d becomes

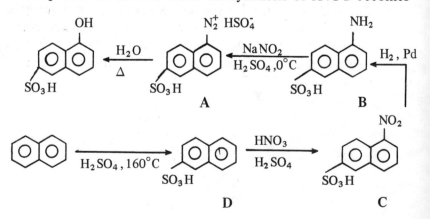

XVI-e

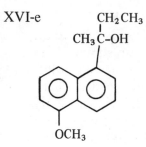

Let us ask some questions concerning XVI-e.

(1) What kind of molecule is XVI-e?
   Answer: XVI-e is a disubstituted naphthalene [2-(5-methoxy-1-naphthyl)-2-butanol].

(2) What procedures are available for placing an –OCH$_3$ group on a naphthalene nucleus?
   Answer: Analogous to the benzene system, an –OCH$_3$ group can be placed on a naphthalene nucleus by treating the corresponding naphthol with CH$_2$N$_2$. The Na salt of the naphthol can also be treated with CH$_3$I to yield the –OCH$_3$ group as shown in procedure VII-1.

(3) What procedure is available for attaching an alcohol to a naphthalene nucleus?
   Answer: The alcohol is a tertiary alcohol, thus it can be prepared from a Grignard reagent and ketone. Also, since the carbinol carbon is attached directly to the aromatic nucleus, the naphthalene carbinol carbon bond is the one which should be cleaved as shown below.

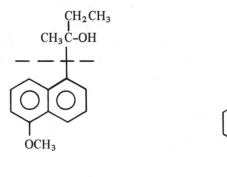
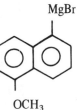

(4) How do we start the synthesis of XIV-e?
Answer: The substitution pattern of XIV-e is important.
The –OCH$_3$ group being an activating group should cause
electrophilic substitution to occur in the same ring, while
a Br would direct electrophilic substitution to the other
ring. It should also be noted that since the alcohol is tertiary it is very susceptible to elimination and thus should
probably be placed on the molecule in the last step. Thus
if we are to run a Grignard reaction in the last step we
need **A**.

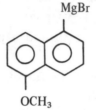

A

(5) How can **A** be prepared?
Answer: By treating **B** with Mg.

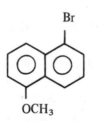

B

(6) How can **B** be prepared?
Answer: The Br is a deactivating group and will direct
electrophilic substitution to the desired position. Thus the
–OCH$_3$ group should be put on the ring next. Let us use
CH$_2$N$_2$ which will require the naphthol **C**.

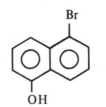

C

(7) How can **C** be prepared?
Answer: By hydrolyzing the diazonium salt **D** via procedure XVI-6.

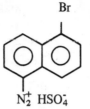

D

(8) How can **D** be prepared?
Answer: By diazotizing the amine **E** via procedure XII-1.

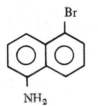

E

(9) How can **E** be prepared?
Answer: By reduction of **F** via procedure XVI-5.

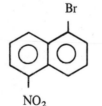

F

(10) How can **F** be prepared?
Answer: By nitration of **G** via procedure XVI-2.

G

(11) How can **G** be prepared?
Answer: By treating naphthalene with Br_2 via procedure XVI-1.

Thus the synthesis of XVI-e becomes

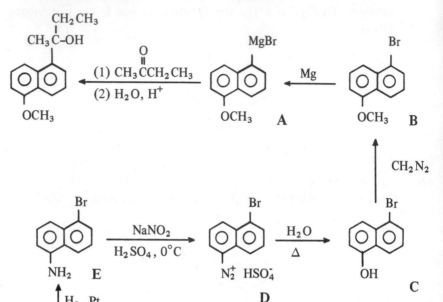

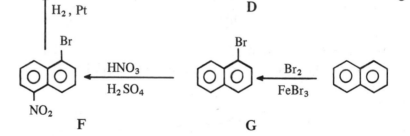

UNWORKED SYNTHESES

Using the same instructions as stated for worked syntheses, prepare the following

(f)

(g)

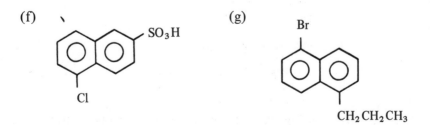

(h)

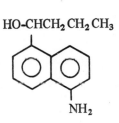

OH

NH_2

(i)

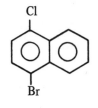

Cl

Br

(j)

$HO-CHCH_2 CH_2 CH_3$

NH_2

(k)

$OCH_2 CH_3$

$SO_2 Cl$

(l)

NO_2 NH_2

(m)

$H-N-\overset{\overset{\displaystyle O}{\|}}{C}CH_3$

Br

(n)

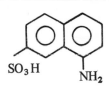

$SO_3 H$ NH_2

(o)

$\overset{\displaystyle CH_2 CH_3}{\underset{\displaystyle |}{HO-CCH_2 CH_3}}$

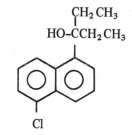

Cl

ANSWER SECTION

h) $CH_3CH_2CH_2CH_2CH_2CH_2CH_3$ $\xleftarrow[(I-3a)]{CH_3CH_2CH_2CH_2Br}$ $(CH_3CH_2CH_2)_2CuLi$

$\Big\uparrow CuI$ $(I-3a)$

$CH_3CH_2CH_2I \xrightarrow[(I-3a)]{Li} CH_3CH_2CH_2Li$

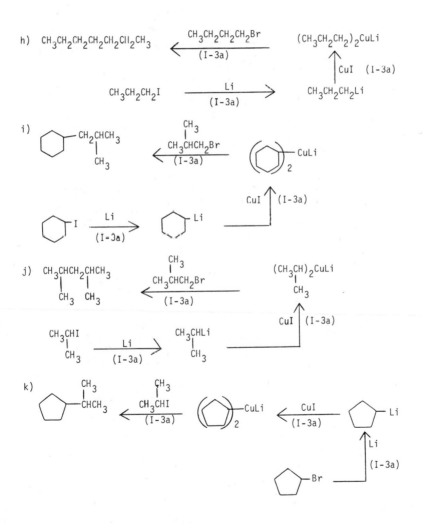

i)

j) $CH_3CHCH_2CHCH_3$

k)

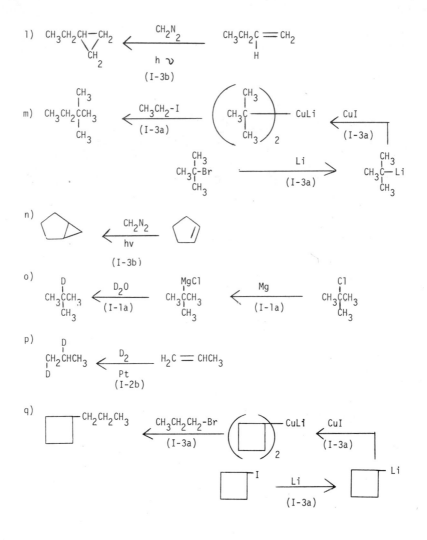

CHAPTER II

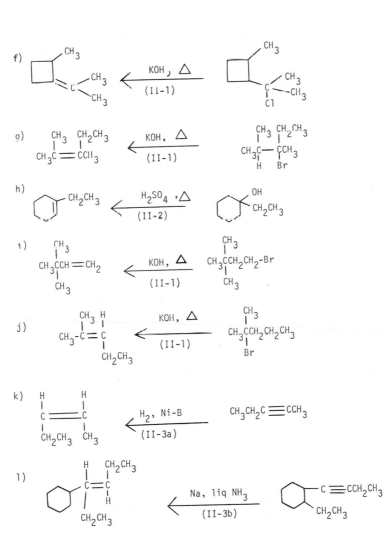

269

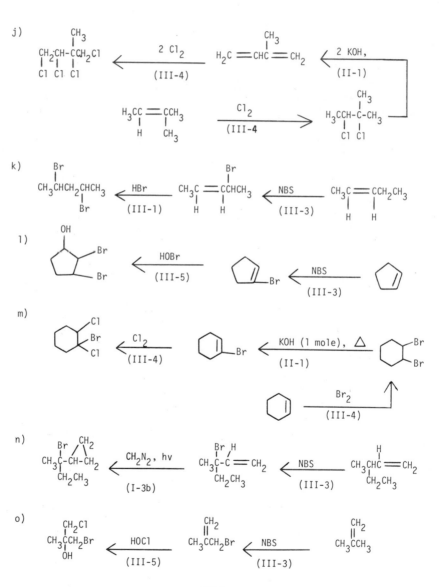

j)

k)

l)

m)

n)

o)

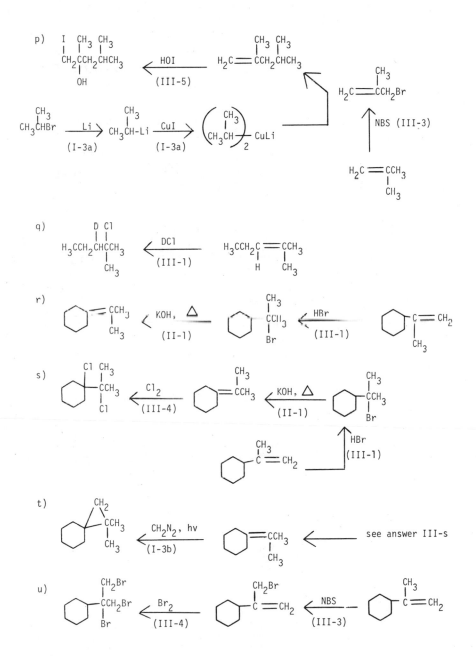

f)

$$CH_3CH_2CH-\underset{\underset{Br}{|}}{\overset{\overset{CH_3}{|}}{C}}-CHBr \quad \xleftarrow[\text{(III-4)}]{2\ Br_2} \quad CH_3CH_2CHC\equiv CH \quad \longleftarrow$$

(with Br, Br on the middle carbon)

$$CH_3CH_2\overset{\overset{CH_3}{|}}{C}HBr$$

(IV-2)

$$HC\equiv CH \quad \xrightarrow[\text{(IV-2)}]{Na,\ liq\ NH_3} \quad HC\equiv C^-\ Na^+$$

g)

$$CH_3CH_2\underset{\underset{Cl}{|}}{\overset{\overset{Cl}{|}}{C}}HCHCH_2CH_3 \quad \xleftarrow[\text{(III-4)}]{Cl_2} \quad CH_3CH_2CH=CHCH_2CH_3 \longleftarrow$$

$$Na,\ liq\ NH_3$$
(II-3b)

$$CH_3CH_2C\equiv CH \quad \xrightarrow[\text{(IV-2)}]{Na,\ liq\ NH_3} \quad CH_3CH_2C\equiv C^-\ Na^+ \quad \xrightarrow[\text{(IV-2)}]{CH_3CH_2Br} \quad CH_3CH_2C\equiv CCH_2CH_3$$

h)

$$CH_3\underset{\underset{D}{|}\ \underset{D}{|}}{\overset{\overset{Br}{|}\ \overset{Br}{|}}{C}}-CCH_2CH_2CH_3 \quad \xleftarrow[\text{(I-2)}]{D_2,\ Pt} \quad CH_3\overset{\overset{Br}{|}}{C}=\overset{\overset{Br}{|}}{C}CH_2CH_2CH_3 \longleftarrow$$

$$Br_2\ (1\ mole)$$
(III-4)

$$\xrightarrow[\text{(IV-2)}]{Na,\ liq\ NH_3} \quad CH_3C\equiv C^-\ Na^+ \quad \xrightarrow[\text{(IV-2)}]{CH_3CH_2CH_2Br} \quad CH_3C\equiv CCH_2CH_2CH_3$$

$$CH_3C\equiv CH$$

i)

$$CH_3\underset{\underset{CH_3}{|}}{C}HCH_2\underset{\underset{Cl}{|}}{\overset{\overset{Cl}{|}}{C}}CH_3 \quad \xleftarrow[\text{(III-1)}]{2\ HCl} \quad CH_3\underset{\underset{CH_3}{|}}{C}HCH_2C\equiv CH \quad \xleftarrow \quad CH_3\underset{\underset{CH_3}{|}}{C}HCH_2Br$$

(IV-2)

$$HC\equiv CH \quad \xrightarrow[\text{(IV-2)}]{Na,\ liq\ NH_3} \quad HC\equiv C^-\ Na^+$$

j)

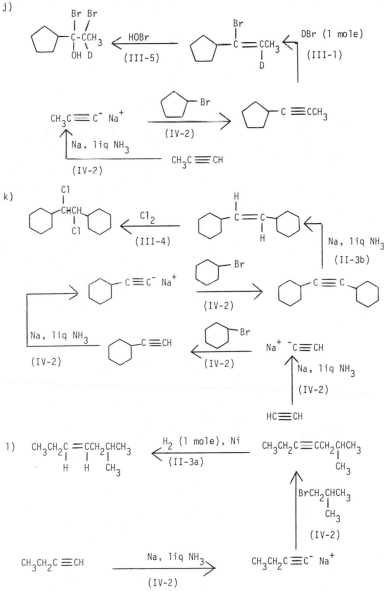

k)

l)

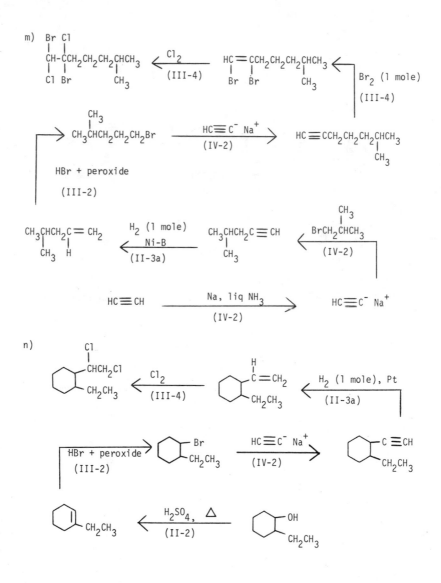

o)

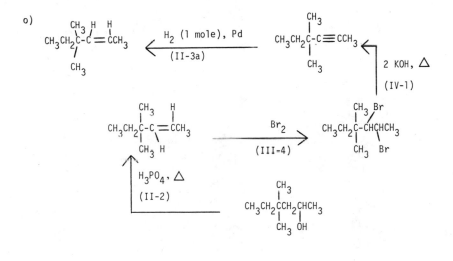

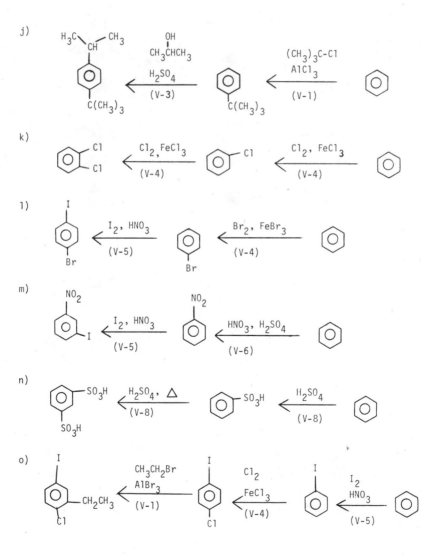

CHAPTER V

276

p)

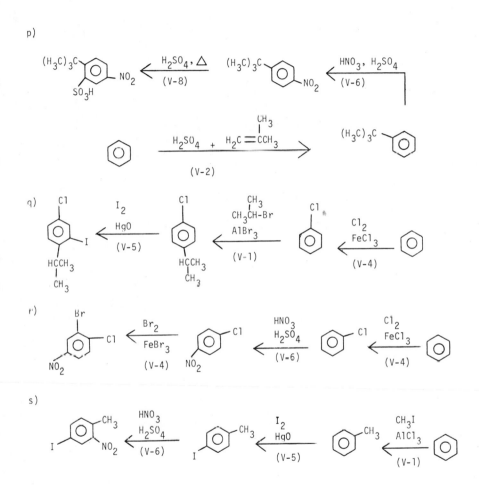

q)

r)

s)

j)

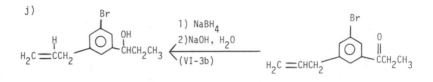

k)

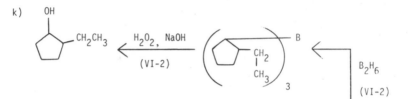

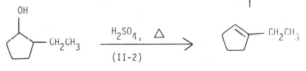

l)

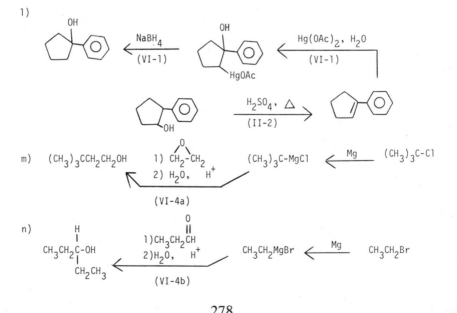

m) $(CH_3)_3CCH_2CH_2OH$ $\begin{array}{c} O \\ \diagup\hspace{-0.3em}\backslash \\ 1)\ CH_2-CH_2 \\ 2)\ H_2O,\quad H^+ \end{array}$ $(CH_3)_3C-MgCl$ $\xleftarrow{\text{Mg}}$ $(CH_3)_3C-Cl$

(VI-4a)

n) $\begin{array}{c} H \\ | \\ CH_3CH_2C-OH \\ | \\ CH_2CH_3 \end{array}$ $\begin{array}{c} O \\ \| \\ 1)\ CH_3CH_2CH \\ 2)\ H_2O,\quad H^+ \end{array}$ CH_3CH_2MgBr $\xleftarrow{\text{Mg}}$ CH_3CH_2Br

(VI-4b)

278

o)

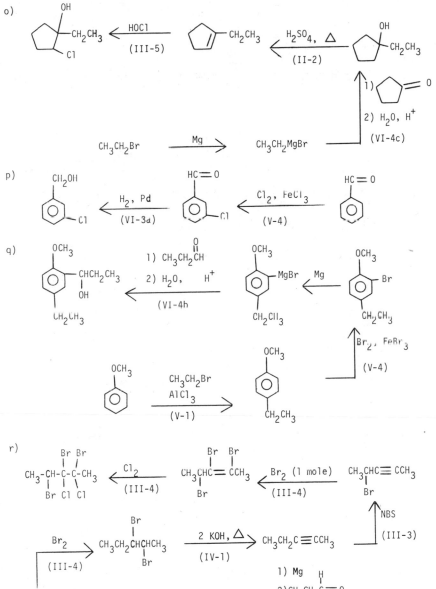

p)

q)

r)

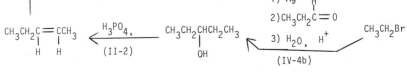

s)

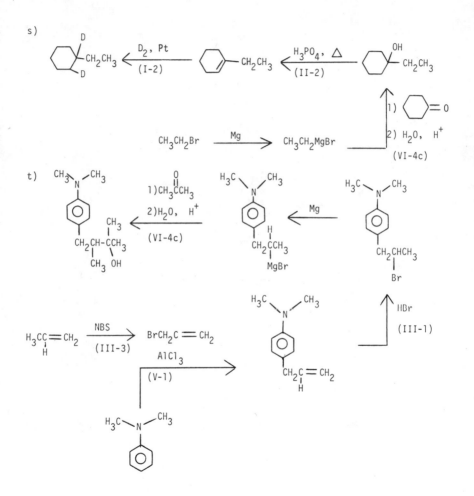

t)

u)

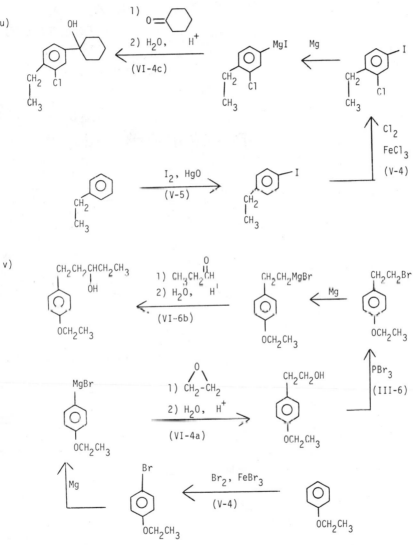

v)

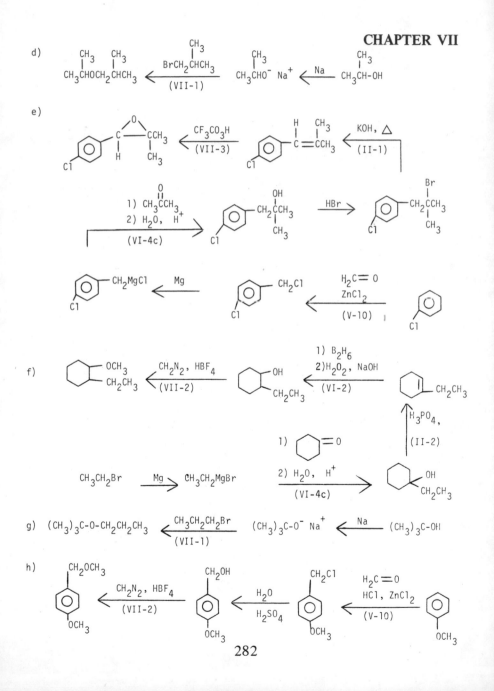

i)

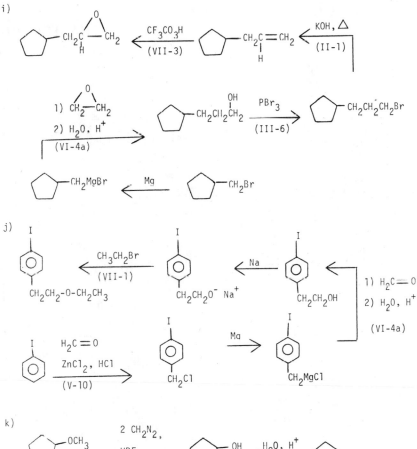

j)

k)

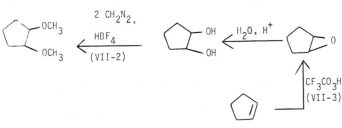

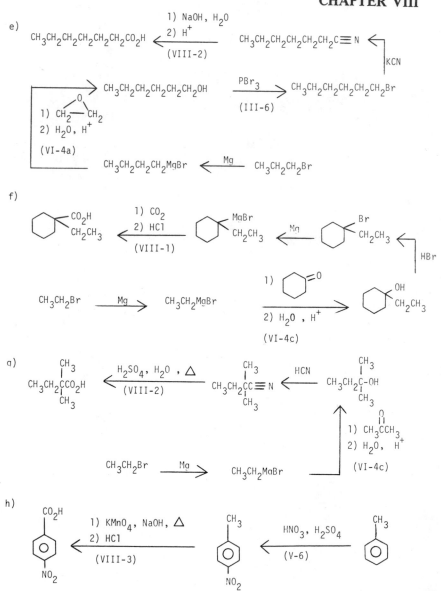

e) $CH_3CH_2CH_2CH_2CH_2CH_2CO_2H$ $\xleftarrow[\text{(VIII-2)}]{\begin{array}{c}1)\ NaOH,\ H_2O\\2)\ H^+\end{array}}$ $CH_3CH_2CH_2CH_2CH_2CH_2C\equiv N$ $\xleftarrow{KCN}$

$\xrightarrow{}$ $CH_3CH_2CH_2CH_2CH_2CH_2OH$ $\xrightarrow{PBr_3}$ $CH_3CH_2CH_2CH_2CH_2CH_2Br$

1) $CH_2\!-\!CH_2$ (O)
2) $H_2O,\ H^+$

(VI-4a)   (III-6)

$CH_3CH_2CH_2CH_2MgBr$ $\xleftarrow{Mg}$ $CH_3CH_2CH_2Br$

f) $\xleftarrow[\text{(VIII-1)}]{\begin{array}{c}1)\ CO_2\\2)\ HCl\end{array}}$ $\xleftarrow{Mg}$ $\xleftarrow{HBr}$

CH_3CH_2Br $\xrightarrow{Mg}$ CH_3CH_2MgBr

$\xrightarrow[\text{2) }H_2O\ ,\ H^+]{1)}$

(VI-4c)

a) $CH_3CH_2\overset{\underset{\displaystyle CH_3}{|}}{\underset{\underset{\displaystyle CH_3}{|}}{C}}CO_2H$ $\xleftarrow[\text{(VIII-2)}]{H_2SO_4,\ H_2O\ ,\ \Delta}$ $CH_3CH_2\overset{\underset{\displaystyle CH_3}{|}}{\underset{\underset{\displaystyle CH_3}{|}}{C}}C\equiv N$ $\xleftarrow{HCN}$ $CH_3CH_2\overset{\underset{\displaystyle CH_3}{|}}{\underset{\underset{\displaystyle CH_3}{|}}{C}}\!-\!OH$

1) $CH_3\overset{O}{\overset{||}{C}}CH_3$
2) $H_2O,\ H^+$

CH_3CH_2Br $\xrightarrow{Mg}$ CH_3CH_2MgBr (VI-4c)

h) $\xleftarrow[\text{(VIII-3)}]{\begin{array}{c}1)\ KMnO_4,\ NaOH,\ \Delta\\2)\ HCl\end{array}}$ $\xleftarrow[\text{(V-6)}]{HNO_3,\ H_2SO_4}$

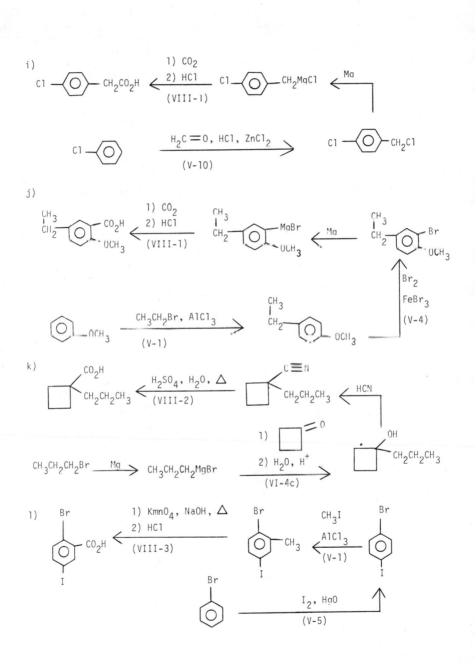

g)

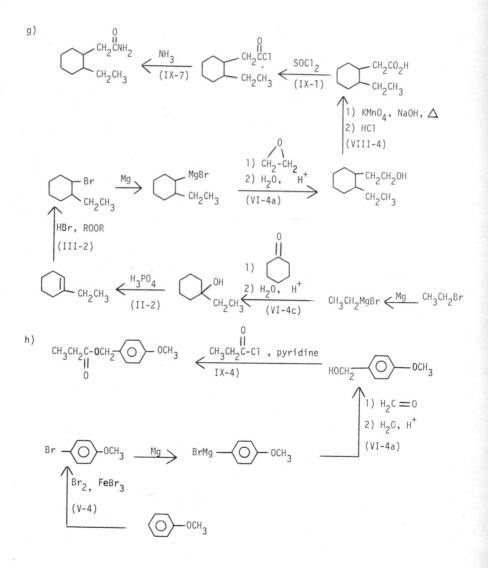

h)

i)

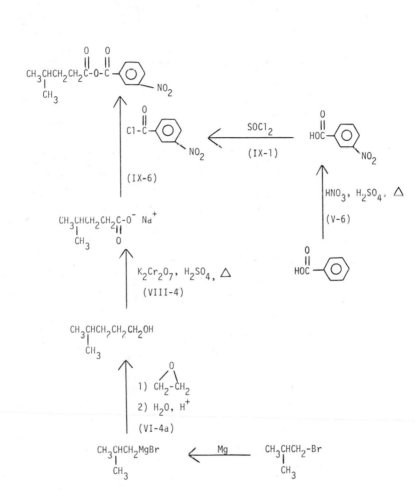

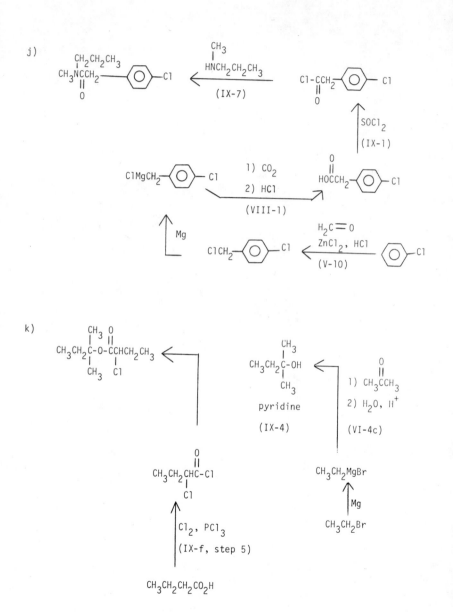

1)
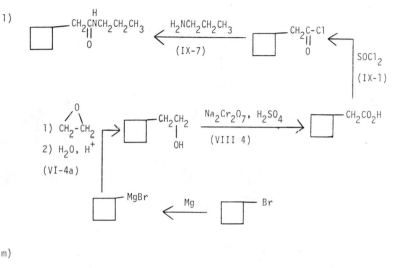

m)
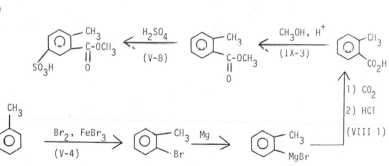

n)

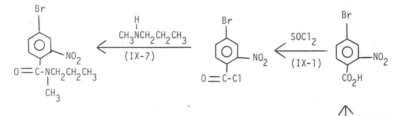

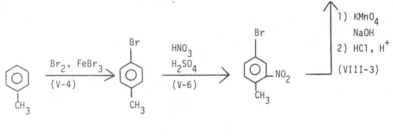

o)

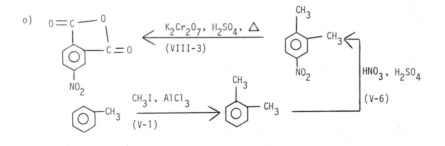

i)

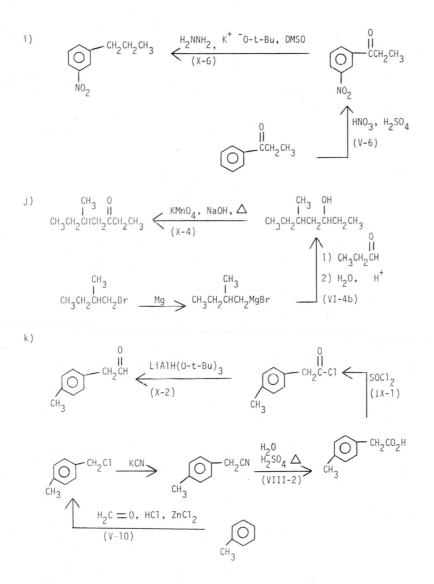

j)

k)

1)

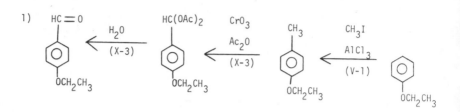

m)

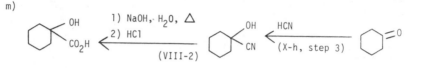

n)

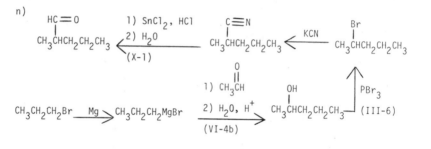

o)

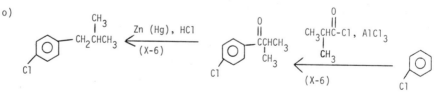

p)

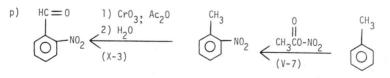

q)

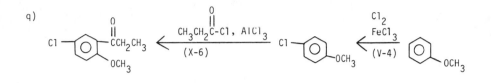

r)

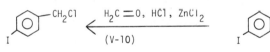

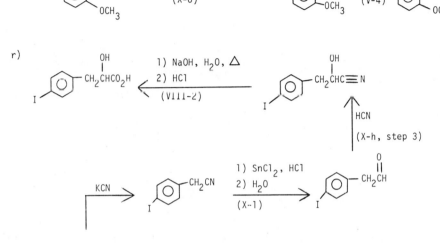

s)

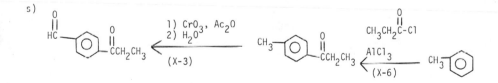

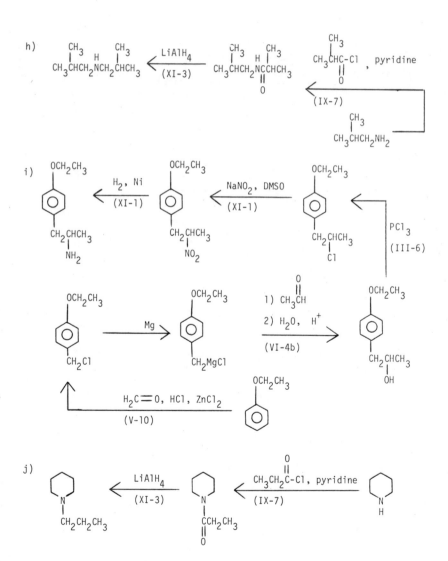

294

k)

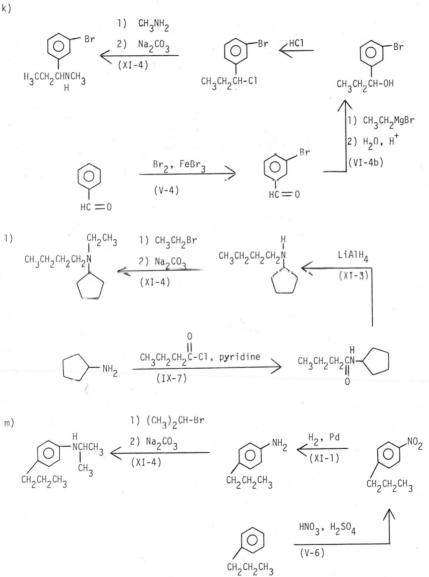

l)

m)

n)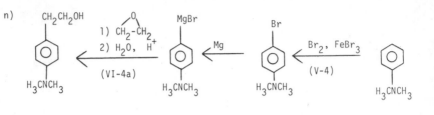

o)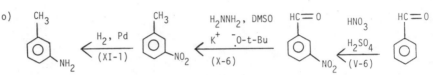

p)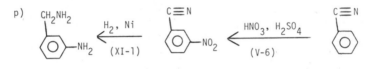

q)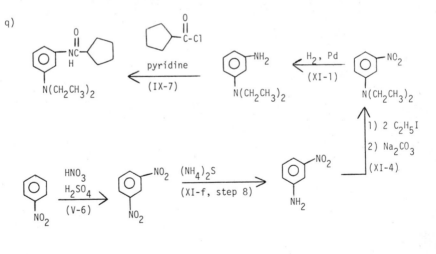

r)

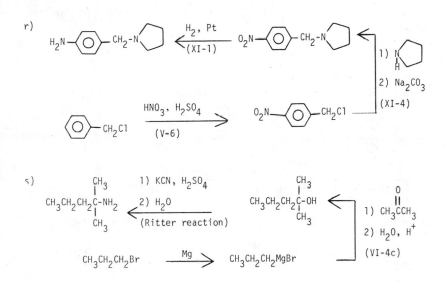

s)

$$CH_3CH_2CH_2\underset{\underset{CH_3}{|}}{\overset{\overset{CH_3}{|}}{C}}-NH_2 \quad \underset{\underset{\text{(Ritter reaction)}}{\xleftarrow{\hspace{1.5cm}}}}{\overset{\text{1) KCN, } H_2SO_4}{\overset{\text{2) } H_2O}{}}}$$

$$CH_3CH_2CH_2\underset{\underset{CH_3}{|}}{\overset{\overset{CH_3}{|}}{C}}-OH$$

1) CH$_3$CCH$_3$

2) H$_2$O, H$^+$

(VI-4c)

$$CH_3CH_2CH_2Br \quad \xrightarrow{\text{Mg}} \quad CH_3CH_2CH_2MgBr$$

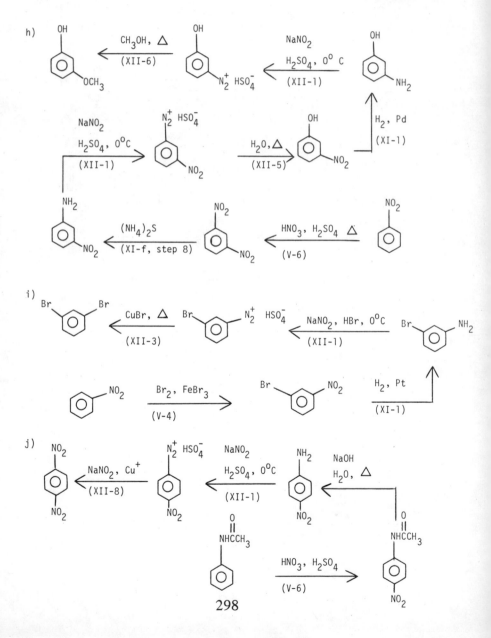

CHAPTER XII

298

k)

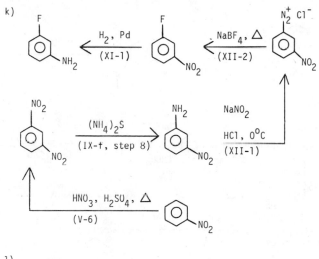

l)

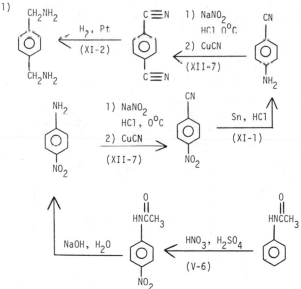

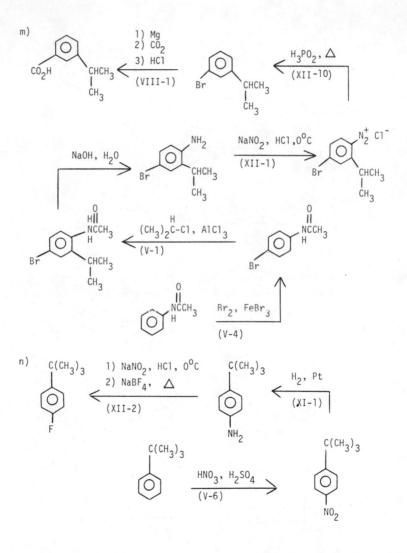

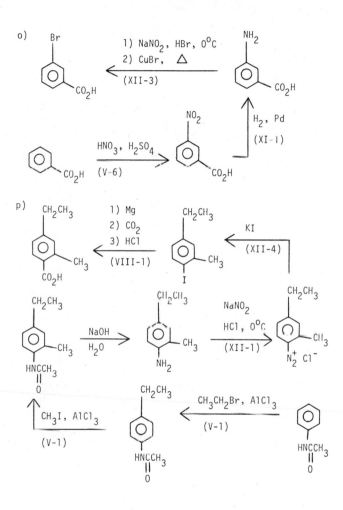

o)

1) NaNO$_2$, HBr, 0°C
2) CuBr, $\triangle$
(XII-3)

NH$_2$

HNO$_3$, H$_2$SO$_4$
(V-6)

H$_2$, Pd
(XI-1)

p)

1) Mg
2) CO$_2$
3) HCl
(VIII-1)

KI
(XII-4)

NaOH
H$_2$O

NaNO$_2$
HCl, 0°C
(XII-1)

CH$_3$I, AlCl$_3$
(V-1)

CH$_3$CH$_2$Br, AlCl$_3$
(V-1)

q)

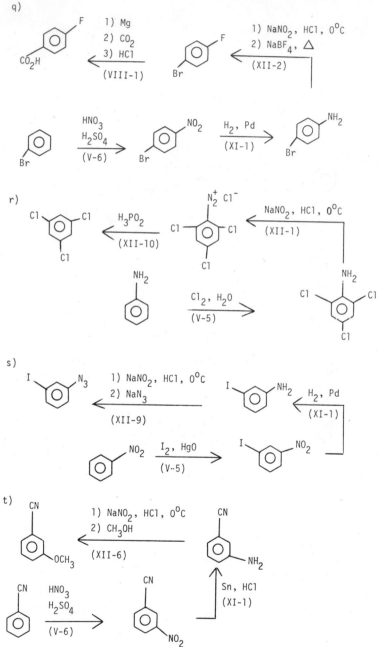

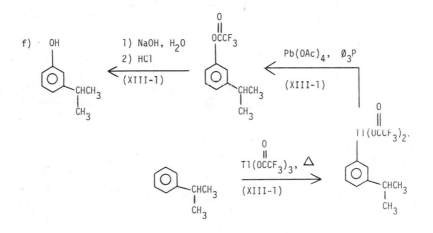

g)

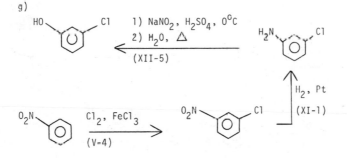

h)

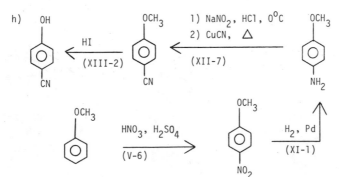

303

i)

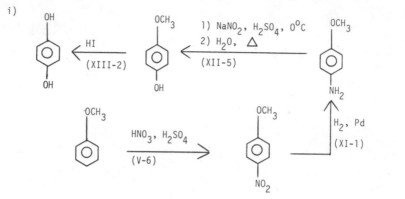

j)

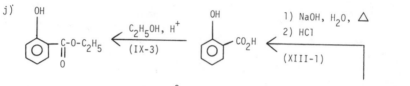

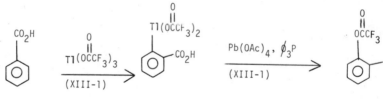

k)

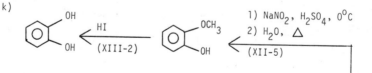

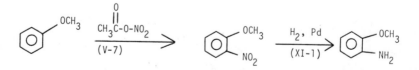

1)

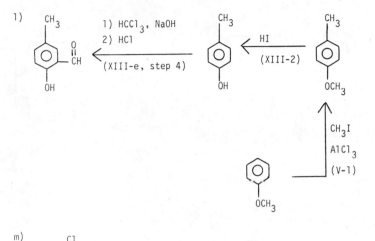

m)

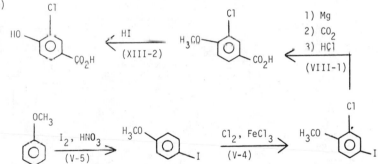

n)

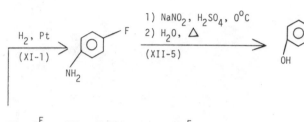

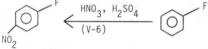

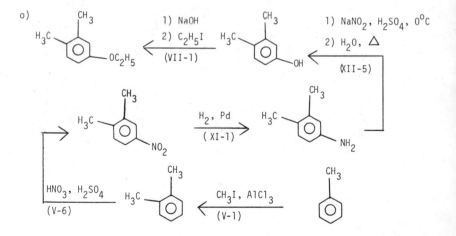

p)

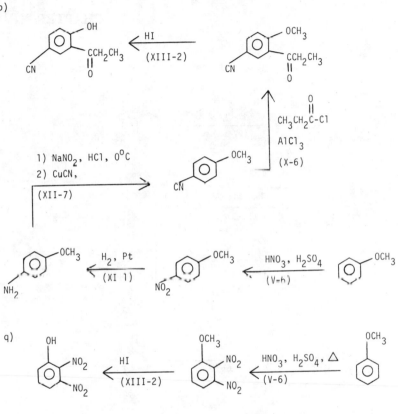

q)

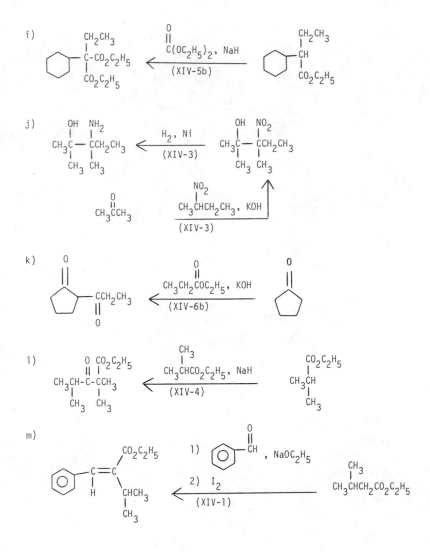

n)

$$CH_3CH_2CHCH-NO_2 \quad \overset{OH}{|} \quad \xleftarrow[\text{(XIV-3)}]{CH_3CH_2\overset{\overset{O}{\|}}{C}H, \ NaOH} \quad CH_3CH_2CH_2NO_2$$

$$\underset{CH_2CH_3}{|}$$

o)

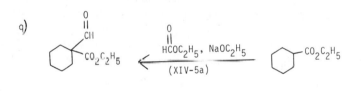

p)

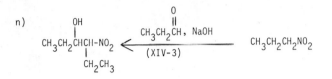

q)

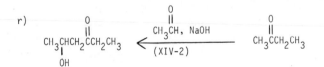

r)

$$\underset{\underset{OH}{|}}{CH_3CHCH_2\overset{\overset{O}{\|}}{C}CH_2CH_3} \quad \xleftarrow[\text{(XIV-2)}]{CH_3\overset{\overset{O}{\|}}{C}H, \ NaOH} \quad CH_3\overset{\overset{O}{\|}}{C}CH_2CH_3$$

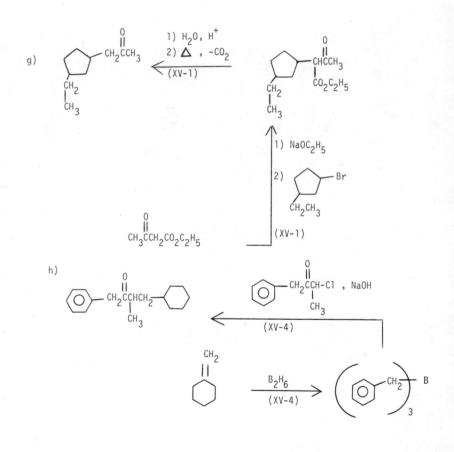

g)

1) H$_2$O, H$^+$
2) Δ , -CO$_2$
(XV-1)

1) NaOC$_2$H$_5$

2)

(XV-1)

CH$_3$CCH$_2$CO$_2$C$_2$H$_5$

h)

-CH$_2$CCHCH$_2$-
CH$_3$

-CH$_2$CCH-Cl , NaOH
CH$_3$

(XV-4)

CH$_2$

B$_2$H$_6$
(XV-4)

-CH$_2$ B

3

310

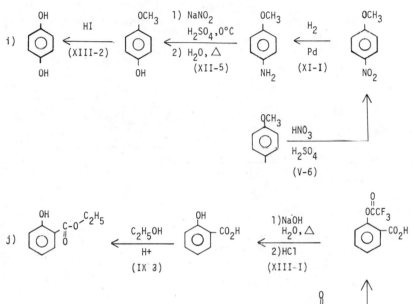

i)

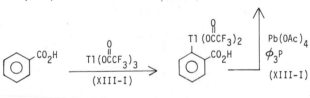

j)

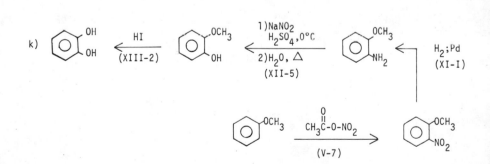

k)

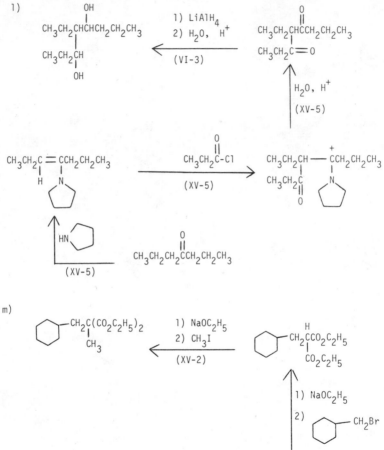

n)

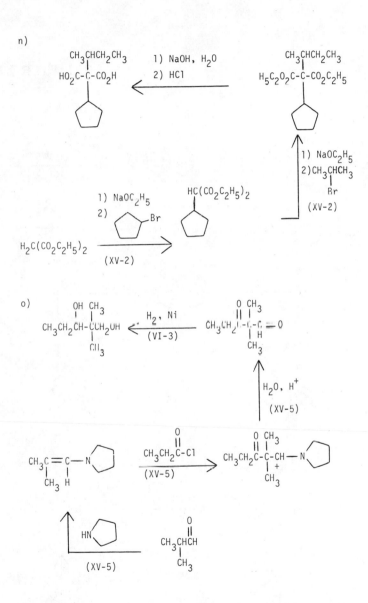

o)

f)

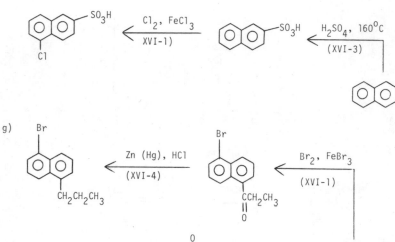

g)

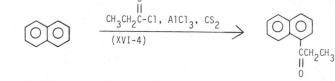

h)

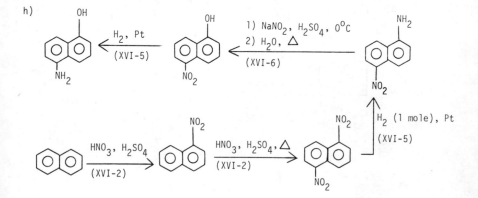

314

i)

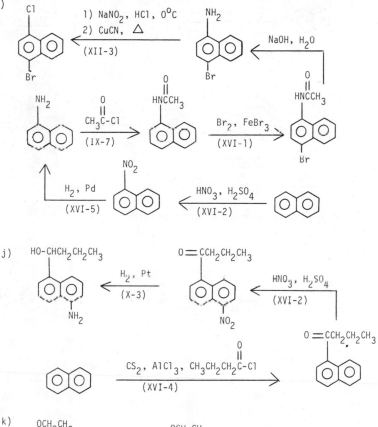

j)

k)

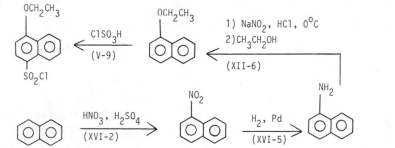

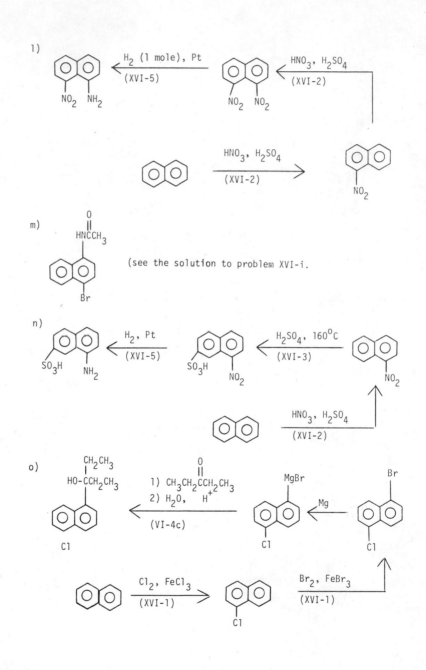

1)

H₂ (1 mole), Pt
(XVI-5)

HNO₃, H₂SO₄
(XVI-2)

HNO₃, H₂SO₄
(XVI-2)

m)

(see the solution to problem XVI-i.

n)

H₂, Pt
(XVI-5)

H₂SO₄, 160°C
(XVI-3)

HNO₃, H₂SO₄
(XVI-2)

o)

1) CH₃CH₂CCH₂CH₃
2) H₂O, H⁺
(VI-4c)

Mg

Cl₂, FeCl₃
(XVI-1)

Br₂, FeBr₃
(XVI-1)

INDEX

In this index, if a synthetic procedure has been introduced in the procedure section of a chapter, it will be indicated by a Roman numeral-number. In a few cases a synthetic procedure was introduced in the worked synthesis portion of a particular problem; this procedure will be indicated by Roman numeral-letter, step number.

Halides - Alkyl

Halides - Aryl

Hydrocarbons - Alkanes

Hydrocarbons - Alkenes

Hydrocarbons - Alkynes

Ketones

NAME REACTIONS